Combined Proceedings International Plant Propagators' Society 2021

IPPS Volume 71

Held 2021

ISBN: 978-1-7138-4505-8

Printed from e-media with permission by:

Curran Associates, Inc.
57 Morehouse Lane
Red Hook, NY 12571

International Plant Propagators' Society (IPPS)
174 Crestview Drive
Bellefonte, PA 16823-8516
USA

Phone: +01 814-355-2467

secretary@ipps.org

Additional copies of this publication are available from:

Curran Associates, Inc.
57 Morehouse Lane
Red Hook, NY 12571 USA
Phone: 845-758-0400
Fax: 845-758-2633
Email: curran@proceedings.com
Web: www.proceedings.com

THE INTERNATIONAL PLANT PROPAGATORS' SOCIETY

Combined Proceedings

Volume 71

2021

IPPS

Sharing Plant Production Knowledge Globally

Published 2021

Combined Proceedings International Plant Propagators' Society (ISSN 0538-9143) is a yearly publication of the International Plant Propagators' Society (IPPS) headquartered in Pennsylvania, U.S.A.
Email: Secretary@ipps.org
Website: www.ipps.org

Business Office. Memberships are processed at the regional level, with dues being allocated to both international and regional expenses. Contact the regional secretary / treasurer in your area of the world for membership information and applications.

Table of Contents

Southern Region - USA

PROCEEDING'S PAPERS

NEW ZEALAND REGION

Jill Reader, Regional Editor

Forty-ninth Annual Meeting - 2021

Hamilton, New Zealand

Combined Proceedings IPPS 71:1-6. 2021.

Plant Variety Rights Review: What is Happening?

Chris Barnaby

Plant Variety Rights Office, Intellectual Property Office of New Zealand, Ministry of Business, Innovation and Employment Private Bag 4714, Christchurch

chris.barnaby@pvr.govt.nz

Keywords: Plant Variety Rights, Treaty of Waitangi, UPOV Convention, scope of protection, propagating material, essential derivation, harvested material

Summary

Plant Variety Rights (PVR) are an intellectual property Right specifically developed for plant breeders, providing a tool for the commercialisation of cultivars and the opportunity to make a return on their investment in developing new plant varieties. The review by Ministry of Business, Innovation and Employment (MBIE) began in 2017 and continued through 2018 with the public release of an issues paper in September 2018. Further consultation occurred with industry and Maori in 2019 and concluded with Cabinet approval for PVR in the Legislation Programme. The draft PVR Bill had the first reading in Parliament in May 2021.

The review has obligations under the Treaty of Waitangi, the 1991 UPOV Convention and the Comprehensive and Progressive Agreement for Trans-Pacific Partnership (CPTPP). The Waitangi Tribunal report for Wai262 has formed the basis of change in the management of applications for taonga species and the 1991 UPOV Convention has provided guidance and recommendations on what is included in the new law including the greater scope of Rights, the addition of Essential Derivation and limited Rights over harvested material.

MBIE, Intellectual Property Office of New Zealand (IPONZ) and Plant Variety Rights Office (PVRO) have also conducted an internal administrative review of Regulations and PVRO practice which has proposed changes in administrative and operational practice.

Obligations under the CPTPP have set a tight timeframe for introduction of the new law with the intention to be in force by end of 2021.

INTRODUCTION

Plant Variety Rights (PVR) are an intellectual property Right specifically developed for plant breeders, providing a tool for the commercialisation of cultivars and the opportunity to make a return on their investment in developing new plant varieties. The existing law, coming into force in 1987, provides for the grant of a fixed term of intellectual property to breeders or owners over their new plant varieties. The exclusive grant of Rights applies to the production for sale and selling of propagating material of new cultivars, but this is now dated, with only a few minor amendments over the last thirty years. Industry has been calling for change for at least twenty years and now finally it is happening.

The current review has two main drivers; the obligations under the Treaty of Waitangi and those of the Comprehensive and Progressive Agreement for Trans -Pacific Partnership (CPTPP).

WHAT HAS OCCURRED?

The review by Ministry of Business, Innovation and Employment (MBIE) began in 2017 and a consultation series of hui with Maori and meetings with industry occurred in 2018. This culminated with the public release of an issues paper in September 2018. Throughout 2019 there was further consultation with Maori, industry and other interested parties in order to develop and draft policy options. The final result was Cabinet approving the addition of PVR to the Legislation Programme in November 2019 and the first reading of the new PVR Bill in Parliament in May 2021.

In parallel to the main policy programme, MBIE, Intellectual Property Office of New Zealand (IPONZ) and Plant Variety Rights Office (PVRO) have conducted an operational and administrative review of Regulations, Fees and PVRO practices. Regulations and Fees will require legislative change with probable implementation in 2022. Changes to or new practices and processes which do not require legislative change will be progressively applied to support the new law.

The Treaty of Waitangi

In the middle 2011, Ko Aotearoa Tenei, the Wai 262 report was released to the public. The report is an extensive document and covered intellectual property and taonga works, genetic and biological resources of taonga species and the environment and matauranga Maori. Although a relatively small component in the context of the full report, PVR is specifically addressed and four recommendations in relation to PVR and taonga species were made.

1. The Commissioner of PVR be empowered to refuse a grant that would affect the kaitiaki relationship;

2. The Commissioner of PVR be supported by a Māori advisory committee;

3. A definition of 'breed' be included to clarify that a plant simply discovered in the wild would not be eligible for a PVR;

4. The Commissioner of PVRs be enabled to refuse a denomination (name) for a new variety if registration or use of that name would offend a significant section of the community including Māori.

The four recommendations are the starting point for meeting the Crown's obligations under the Treaty.

The second recommendation involves the establishment of a Maori Committee (MC) to work alongside the PVR Commissioner to ensure that kaitiaki relationships with taonga species are adequately considered with respect to PVR applications. The management of kaitiaki relationships is a key recommendation from the Wai 262 report. Engagement with Māori during the review period highlighted the importance of kaitiaki being involved with breeders of taonga species at an early stage, before any PVR application is made. The proposed Māori Committee should play a role in achieving this, fostering partnerships between native plant breeders and local Maori. At this time, the details of the way the Māori Committee will function have not been settled, but it is clear that the primary objective will be to ensure improved consideration for taonga species. The intention is for the Maori Committee to have responsibility for meeting all Treaty elements of any new law.

At this stage a formal definition of taonga species is not available however the species included in this group are likely to be all native or indigenous plants and a very limited number of others such as kumara (*Ipomea batatas*).

In a broader sense, there is uncertainty regarding how the proposed changes will practically impact PVR for taonga species. The proposals outline the intention for kaitiaki relationships to be acknowledged and addressed for taonga species which will require breeders using these species to engage with Maori as part of breeding actions and understand that the PVR application process will include submission of the variety to the Maori Committee.

Using application data from recent years it is estimated that around 7% of applications are belonging to taonga species, in the order of 7-11 varieties per year will be required to be submitted to the MC. All other varieties, over 90% of applications, will not be submitted to the MC and there will be no Treaty of Waitangi provisions applied to applications for those varieties.

Convention of the International Union for the Protection of New Varieties of Plants (UPOV)

Plant Variety Protection legislation in most countries is based on either the 1978 or 1991 Convention. The Convention consists of a series of Articles which list the requirements for national law compliance. At present, New Zealand's law is aligned with the 1978 Convention. The majority of UPOV member states are aligned with the 1991 Convention. To meet obligations under CPTPP trade agreement, the NZ PVR regime needs to be upgraded and give effect to 1991.

The 1991 Convention provides for stronger Rights, with the following of most significance:

1. Greater scope of protection
2. Essential Derivation
3. Rights over harvested material
4. Exception provisions for farmer saving of seed

SCOPE OF PROTECTION

The scope of protection has been expanded from a focus on commercial propagation and the sale of propagating material including whole plants to a much broader objective of commercialisation or exploitation of the variety as a whole. The existing Rights over commercial propagation, reproduction and multiplication are retained and continue to encompass offering for sale, selling and marketing of plants of the variety. The scope has been extended to include; conditioning for propagation, exporting, importing and stocking for any of these activities. All of these activities will now require the permission of the breeder.

An example of how the new law could make management of a Right easier is the situation where a breeder becomes aware of a nursery stocking one of the breeder's varieties. Under the current law the breeder would have to establish that the nursery was actually propagating and selling the variety in order to take infringement action. Under the new law, the presence or stocking of plants of the variety alone may be sufficient to initiate infringement action. In any infringement situation it is recommended to obtain professional legal advice.

A second example is the unauthorised export of plant material of a protected variety. Currently it is very difficult for export to be prevented because the variety owner would have to establish that commercial propagation of that material had occurred. Not an easy thing to do. The new law will remove the need to establish commercial activity because the export itself is an infringement and who propagated the material and how sold becomes a secondary matter.

ESSENTIAL DERIVATION

This provision is an entirely new concept nationally and there is no current equivalent. The concept of one variety being essentially derived from another had its origins in genetic engineering and the concern that a commercially successful variety could be genetically engineered to create a different variety but remain genetically very similar to the initial variety. One variety being genetically similar to another is not confined to genetic modification and could include in bred lines, repeated back crossing and sports (mutations). To address concerns, essential derivation provides the owner of a protected initial variety the possibility to share in the commercialisation of any other variety predominantly derived from that original variety. The derived variety must be distinct from the initial variety and can be protected.

Essential derivation is something of a balance between the important provision that protected varieties are freely available for further breeding and that of the second breeder acknowledging the contribution of the first variety to the second variety. The greatest challenge to Essential Derivation is the definition of a derived variety and how that determination is made. These aspects remain under international discussion and debate and currently there is some variation as to how Essential Derivation in individual national law, is defined and interpreted within the Article in the 1991 UPOV Convention.

HARVESTED MATERIAL

Current PVR law is focused on commercial propagation activity and makes no specific provision for assertion of Rights over harvested material. Harvested material could include fruits, vegetables, cut flowers or grain. The 1991 Convention provides for the owner of a protected variety to have the possibility of asserting their Rights over harvested material, including entire or parts of plants, where there has been unauthorised use of propagating material. This can only be applied where the owner has been unable to assert their Rights at the propagation stage. This provision does not provide a choice for a breeder on when to assert Rights because the assertion of Rights over harvested material is not acceptable if this could have been achieved at the propagation stage.

An example may be where the owner of a pineapple variety protects the variety in New Zealand and then uses that Right to manage the importation of fruit of that variety from a Pacific Island nation. The owner may assert their Rights in New Zealand on the imported fruit because the Pacific Island may not have a PVR scheme and the owner was unable to do this at the time of propagation.

FARM SAVED SEED

Farmers have traditionally freely saved seed for centuries and for crops such as cereals is an important source of seed for planting in the next season. The greater scope of Rights does mean that stocking seed with the intention of future sowing is no longer possible and requires an optional exemption from the Right. This optional exemption will provide for this practice to continue, but variety owners may have a mechanism which could provide for the possibility of asserting their Right over the saved seed. The details of the mechanism have yet to be finalised.

POST GRANT ACTIVITY

Post grant activity includes compulsory licences and infringement matters. Changes have been proposed in all areas with compulsory licences the most contentious.

The purpose of a compulsory licence provision is to ensure that a protected variety of potential or known value to the public cannot be locked up and encourages commercialisation and public benefit. Compulsory licences can be applied after a certain period to any variety that is deemed to be not sufficiently available and the public is missing out. An applicant can apply to the PVR Commissioner and if certain criteria are met, including assessment of public good, could potentially receive a licence which would require the variety owner to make propagating material available to the applicant under terms and conditions set by the PVR Commissioner.

Infringements have also been under scrutiny with industry expressing dissatisfaction with the current Act which provides no guidance for infringement actions as seen in other IP legislation. In common with other intellectual property regimes, the onus is on the variety owner to assert their Right and use civil action when they think their rights have been infringed. The cost of Right enforcement is an issue, in common with other intellectual property types. New legislation could include explanation on the grounds for infringement and remedies that may be considered.

ADMINISTRATIVE AND OPERATIONAL REVIEW

In parallel to the primary policy changes, a review has been carried out of PVRO operational practices and procedures, in particular, practices which are more frequently encountered by users and are more likely to have familiarity with. Examples are organisation of growing trials for DUS testing and timetable for payment of trial or examination fees. In most respects, the user interaction with the online system and PVRO itself is intended to continue largely as it is now. The improvements identified by the review fall into two groups, ones that impact users directly or those that are more for PVRO internal processes. For the future an applicant will likely notice changes in areas such as:

i. Contact address requirements
ii. Time limits for the payment of trial or examination fees
iii. Photo requirements for vegetable and potato varieties
iv. Changes to the requesting of plant material and organisation of testing
v. Website improvements for an improved user-friendly experience

A particular area of consideration has been arrangements for and organisation of growing trials for DUS testing. The current law reflects how things were done in the 1980's and does not fit well the situation of today with greater application numbers and for most genera significantly more potential similar varieties for evaluation. Access to plant material of similar varieties has become increasingly challenging over the last decade due to variety commercialisation models which integrate propagation, production and sales. For varieties managed in this way, the only source of propagating material is often the owner or licensed agent.

Future Timetable

With the Bill in Parliament, the aim is to have a new law in force by end of 2021, with new Regulations in operation by the middle of 2022. The timetable cannot be altered due to the CPTPP requirements that all necessary law be in place within three years of the New Zealand signing in December 2018.

Cost of protection

The PVR scheme is almost entirely funded by users and only receives Government funding for the cost of UPOV membership. All other operational activities are intended to be covered by fees under cost recovery. The overall Act review will include an evaluation of revenue through fees and PVRO expenditure. Such an assessment was last conducted in the early 2000's and is arguable long overdue. Further information regarding the fees review will be available later in 2021.

FURTHER INFORMATION

More information on Plant Variety Rights in New Zealand may be found at the following links:

Plant Variety Rights (IPONZ website): https://www.iponz.govt.nz/about-ip/pvr/

The Plant Variety Rights Act review: https://www.iponz.govt.nz/about-ip/pvr/pvr-act-review/

Combined Proceedings IPPS 71:7-13. 2021.

Measure to Manage – Lab Testing for Growers

Fiona Calvert

Market Sector Manager - Agriculture, Hill Laboratories, Hamilton, New Zealand

fiona.calvert@hill-labs.co.nz

Keywords: Soil testing, substrates, potting media, fertilizer, hydroponics, foliar analysis, water testing

Summary

There are multiple laboratory test options available to nurseries, turf managers and plant growers to help manage their unique production systems. Most of these tests are complementary, in that they each provide a piece of valuable information that may not be entirely useful on its own but when used in conjunction can give confidence in any management change practices, and help with environmental stewardship. Lab testing can also help direct management decisions that may be needed to deal with climate change challenges and regional or national regulation. The sampling protocol used is critical, to guide interpretation and to realise the benefit of testing. Using an expert advisor is highly recommended. Field observations supported by test result data provides a framework to guide decisions, removing guesswork and uncertainty.

INTRODUCTION

Successful growing systems require water, light, heat, air and adequate amounts of essential nutrients throughout the growing cycle of the crop. As well, growers want to protect the environment they live in and be sustainable through an ever-changing world of climate uncertainty and regulation. Laboratory testing can play an important role in achieving successful growing and production systems to manage at least some of these critical elements. Having real data provides assurance to the grower around management decisions and can be useful to help diagnose problems. Lab testing is a risk-management tool to provide confidence of crop supply (in-time), prevent crop losses and to minimize nutrient losses to waterways as well as to protect soil health. Testing is often used for monitoring or else for diagnostic purposes. This paper describes some of the tests available for growers and has commentary on how these tests can be helpful along with some limitations.

Soil Testing

Soil samples are usually collected from the growing zone of plants, and sent to the laboratory to measure the plant-available nutrients. Several samples should be collected from a growing area, as there is often large spatial variability to consider. The bulked sub-samples from an area are analysed as a single sample to provide an average nutrient status for that area. Different crops need to be sampled separately, as they may have different pH and nutrient needs.

In New Zealand the standard tests carried out are pH (water); Olsen Phosphorus, Exchangeable Cations (Ca,K,Mg,Na); Cation Exchange Capacity (CEC); Base Saturation; Volume Weight and sometimes available sulphate-S (SO4) and Soluble Salts (EC). Additional soil tests may include Boron and other Trace Elements where special investigation are needed.

Testing for Organic Matter, Total Carbon, Total Nitrogen, the CN ratio, Mineralisable Nitrogen and Hot Water Extractable Carbon are also carried out routinely, as growers look to monitor soil health using easy tests on the sample already submitted for fertility monitoring.

The soil pH test can help with understanding nutrient availability, as shown for a mineral soil diagrammatically below. Fig 1 is for a mineral soil and organic soils behave a little differently.

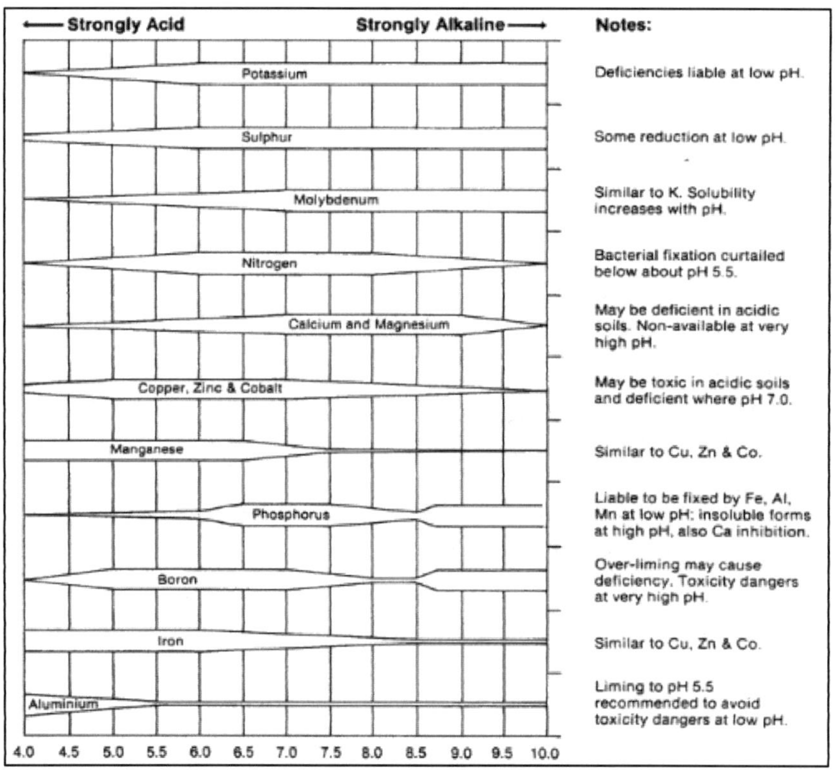

Figure 1. Soil pH effects on plant availability of nutrients (Truog, 1948)

A soil pH range of 5.8-6.5 is considered ideal for most plants (lower range for organic soils) although some plant species prefer more acid soils e.g. potatoes, camellia. A soil pH near to 6 is considered desirable for earthworm populations to thrive.

Acidification of soil is a natural process under growing plants. As cations are taken up, H^+ ions are released and other processes such as microbial respiration, the release of organic acids by plant roots and use of NH4-N fertilisers all contribute to acidification.

Soil pH can be altered by use of appropriate rate and form of liming product, but it should be noted that plants may grow well in soil or media that doesn't fit the ideal range – provided adequate nutrients are supplied directly to the plant and the plant is reasonably pH-tolerant.

Soil tests can be variable as referred to earlier, both spatially and temporally (over season and years). Sampling consistently from an area and depth will help to minimize variability, but not remove it entirely. Table 1 describes how nutrients in an uncultivated soil may vary in the soil profile with most of the valuable plant-nutrients in the top few centimeters. This will of course be different in soils that are regularly turned over, but highlights the need for sampling at a consistent depth (often 0-150mm for most crops, 0-75mm for turfs).

Table 1. Example of change in nutrient with soil depth (uncultivated soil) from an in-house study by Hill Laboratories.

Soil Depth	pH	Olsen P (mg/L)	Calcium (MAF Units)
0-1"	7.0	60	20
1-2"	6.2	24	10
2-3"	5.8	9	6
3-4"	5.7	7	4
4-5"	5.6	4	3
5-6"	5.6	2	3
Effect of Sampling Depth			
0-3" (0-75mm)	**6.3**	**31**	**12**
0-6" (0-150mm)	**5.9**	**18**	**8**

Growing Media Testing

Although there are some similarities with soil analysis, testing potting media (media) presents some special challenges. In the early 1970s, a 1:1.5 media to water extraction was developed and remains as the standard method for New Zealand and Australia (refer Australian Standard for Potting Mixes AS3743). Testing potting media provides an analysis of immediately available nutrient as well as pH, and is a snapshot in time. Typically, tests carried out from this water extract are pH and Electrical Conductivity (EC), and the immediately available nutrients ammonium-nitrogen, nitrate-nitrogen, phosphorus, sulphur, potassium, calcium, magnesium and sodium.

As for soil tests, the pH of media is important for the availability of nutrients. The EC provides an overall measurement of the dissolved salts, and is important in diagnosing problems such as salt stress in crops.

It is important to realise that a media test measures the nutrients that are immediately plant available. It does not include nutrients that may become available over time (e.g. from slow release fertiliser prills). Consequently it is possible that a Potting Media analysis shows only low levels of nutrients present, even though the crop is apparently growing well. When this is the case, the crop is taking up these nutrients at the same rate as they are being released by slow-release fertilizers.

Heavy watering/leaching of the media just prior to testing may also result in low nutrient levels. The treatment given to the media before or after analysis must be taken into account when interpreting the analysis results. If diagnosing a specific problem, select the sample from the pots showing the most prominent symptoms.

The trace elements iron, manganese, zinc, copper and boron are analysed from a DTPA (Diethylenetriamine pentaacetate) extraction. As with soil trace element testing, there are limitations to the reliability of the test, and in most instances suspected trace element problems should be confirmed with plant tissue analysis.

Desirable pH, soluble salts and nutrient levels vary with the glasshouse crop being grown and management practices. General guidelines are suggested in the table below:

Table2. Table showing general analysis guidelines for potting media of different types (as used for different growing rates).

	Low	**Fairly Low**	**Broad Range**	**Medium**	**High**
pH	5.2 - 6.5	5.2 - 6.5	5.2 - 6.5	5.2 - 6.5	5.2 - 6.5
Conductivity (mS/cm)	0.3 - 0.8	0.5 - 1.2	0.5 - 1.8	1.0 - 1.8	1.0 - 2.5
Nitrate-N (mg/l)	15 - 35	20 - 50	20 - 80	40 - 80	40 - 120
Ammonium-N (mg/l)	1 - 10	1 - 15	1 - 20	3 - 20	1 - 30
Phosphorus (mg/l)	4 - 15	5 - 15	5 - 20	10 - 20	10 - 30
Potassium (mg/l)	10 - 35	20 - 50	20 - 80	40 - 80	40 - 120
Calcium (mg/l)	10 - 20	15 - 40	30 - 70	30 - 70	30 - 100
Magnesium (mg/l)	8 - 15	6 - 15	7 - 25	12 - 25	12 - 35
Sodium (mg/l)	3 - 25	5 - 30	5 - 40	10 - 40	5 - 10

Compost Testing

Compost is most useful in growing systems as a soil amendment, to improve water holding capacity and as a supply of slow-release nutrients. As plants are not usually grown directly in compost, an extractable or plant-available nutrient test is of limited use, apart from pH and conductivity (EC). Analysis of compost is generally carried out using a hot acid digest to measure the total recoverable nutrients along with a Carbon to Nitrogen (CN) ratio, measured by combustion analyser. The CN ratio is important as an indicator of the rate of compost decomposition, and will assist with nitrogen fertiliser decisions. Compost with a high CN ratio (>30) may immobilize soil nitrogen and create a plant "nitrogen-hunger".

Interpretation of compost analysis can be difficult, due to the variety of raw materials in the starting ingredients, but a test can be helpful where heavy metals or other contaminants are suspected. In New Zealand, laboratory methods should align closely with the NZ Standard 4454 - Composts, Soil Conditioners and Mulches.

Plant Testing

Plant testing (also known as plant analysis or tissue testing) is a useful tool for growers, often carried out as complementary to soil or media testing. An analysis should include all of the major and trace elements required by growing plants. Methods used are most often hot acid digestion followed by inductively-coupled-plasma optical emission spectroscopy (ICP-OES) and also combustion analyzer for nitrogen. Extractable methods for chloride and nitrate-N may also be used.

Interpretation can be difficult, as many factors affect the nutrient content of plants, with some of these listed below.

Plant uptake of nutrients may be affected by variables such as:

- Moisture
- Temperature
- Air
- Soil/Media pH
- Physical conditions of the growing media
- Light
- Salinity
- Disease/Pests
- Root Injury
- Residues, toxins

Plant analysis for nutrient content can be a successful tool if an appropriate sampling protocol is used to align with available interpretive ranges. In fact plant sampling is critical, as stage of growth, plant part and plant variety will all impact on interpretation of nutrient levels. One of the reasons for this is to do with nutrient mobility within the plant. *Mobile* nutrients such as nitrogen, phosphorus, magnesium and potassium are redistributed from older leaves to the growing points so that deficiency will be apparent in the older leaves first. *Immobile* nutrients such as calcium, boron and zinc are incorporated into older plant tissue and are not available for subsequent redistribution within the plant. Deficiency of these elements will therefore appear in the growing points of the plant first.

Most reputable laboratories will be able to provide guidance on how to sample, to align with any interpretive ranges in their library. Unique plant varieties should be sampled separately, and most often this is done in a period of active growth. In most cases the sampling part is the youngest mature leaf (YML) to avoid some of the nutrient mobility differences described earlier. Some conventions also suggest sampling the plant petiole only, and this is particularly valid to do for the mobile nutrients.

An alternative approach, especially when there are no reference ranges available, is to sample plants that appear to be in good condition, compared to their same-stage neighbours growing poorly in apparently similar conditions. A "good" versus "bad" set of samples can be helpful, especially when non-nutritional factors are suspected.

Nutrient Solution testing (Hydroponics)

Drip type systems are sampled to compare the feed solution with the drainage solution. Nutrient Film Technique (NFT) and Ebb and Flow systems are sampled to determine the nutrient composition and any changes in the balance of nutrients that occur in the recycled solution.

Interpretation of a nutrient solution analysis is influenced by the choice of crop grown, environmental variables such as light and temperature, the type of hydroponic growing system and how the sample is taken. A typical hydroponic solution test will include the following tests:

- pH as a measure of the acidity (and alkalinity) of the nutrient solution. This should be appropriate for the plants grown and also suitable for maintenance of the solubility of all nutrient components.
- Electrical Conductivity representing the total concentration of dissolved salts (includes nutrients) and is commonly reported as the Conductivity Factor (CF) units.
- Nitrate-Nitrogen and other elements are reported as concentrations of elemental equivalents reported as mg/l which is 'parts per million' on a weight/volume basis.
- The sum of anions and sum of cations which is a comparison of the chemical equivalents of negatively charged ions (anions including P, S, NO3, Cl) with positively charged ions (cations including K, Ca, Mg, Na). These two sums should be similar if the analysis conducted has reported the entire major element components, as the components of a nutrient solution should theoretically be a balance of cations and anions.

Measurements of ammonium-N, molybdenum and silica may also be useful for particular investigative purposes.

Water Testing

The pH, alkalinity, hardness, salts (sodium and chloride) and trace element (boron, iron, manganese) content of the feed water should be researched before commitment to using it in any hydroponic growing system. The simple rule is that the best results are obtained from using 'pure' water. Any dissolved impurities in the water should be present at levels that are lower than the nutrient solution specifications. Treatment options are available to reduce levels of potentially insoluble iron and manganese and to correct pH, alkalinity and hardness. Impurities that are most difficult to manage are dissolved sodium, chloride and boron.

Town water supplies are not always appropriate for hydroponic growing systems without further treatment. Where surface water (river, lake) is used, investigate the risk of contamination from all possible sources within the catchment, such as herbicide applications.

Water used to irrigate container or soil-grown plants may also be less than ideal, depending on its source and other environmental effects. A simple analysis can fore-warn of any potential problems and give confidence in an investment dependent upon supplies of clean water.

Chemical factors affecting the suitability of a water supply for irrigation relate to the presence of a number of potentially undesirable or hazardous features. These include:

• dissolved salts
• chemical toxicities
• the level of sodium and its effect on soil structure

In New Zealand there are several public data sources available, to view regional or local trends in water quality from various water-quality monitoring programs.

Useful websites can be found at:

- LAWA website (Land, Air, Water Aotearoa) www.lawa.org.nz
- Regional Councils – state of the environment reporting e.g. www.waikatoregion.govt.nz/environment (river & stream monitoring)
- MPI website e.g. www.mpi.govt.nz/agriculture (farm management/waterways)

Sources

The information in this article is largely sourced from Hill Laboratories proprietary technical notes and crop guides, available on their website www.hill-laboratories.com

Literature Cited:

Handreck, K.A. and Black, N.D. (eds). (1989). Growing Media for Ornamental Plants & Turf. New South Wales University Press.

Clarke, C.J., Smith, G.S., Prasad, M. and Cornforth, I.S. (1986). Fertiliser Recommendations for horticultural crops. MAF, Wellington.

Cresswell, G.C. and Weir, R.G. (eds). (1997). Plant Nutrient Disorders (5) – Ornamental Plants and Shrubs. NSW Agriculture, Inkata Press.

Cooper, A. (1982). Nutrient Film Technique. The English Language Book Society and Grower Books, London.

Truog, E. (1948). Lime in relation to availability of plant nutrient. Soil Science *65*: 1-7.

Combined Proceedings IPPS 71:14-16. 2021.

Tissue Culture Propagation of Some Temperate Woody Ornamentals

Puthiyaparambil Josekutty

Research Manager, Skybury Coffee Pty Ltd, QLD 4880. www.skybury.com.au; Tel: +61 478793110

josekutty964@gmail.com

Keywords: Micropropagation, redbud, *Cercis*, birch, *Betula*

Summary

Plant tissue culture is a technique of growing isolated plant parts and tissues in aseptic condition on a chemically defined medium under controlled conditions of light, temperature, and humidity. Mass propagation of many woody plants and trees that are difficult to propagate through usual cutting production methods in nursery can be better multiplied in tissue culture. Tissue culture not only offers rapid multiplication but also generates disease free clones. In vitro propagation also allows uninterrupted propagation of plants even in peak winter when the temperate plants generally embrace dormancy. Common tissue culture methods applied for mass propagation include shoot tip / nodal cultures; direct or indirect organogenesis and meristem culture. In this paper, tissue culture propagation of woody ornamentals such as redbud (*Cercis*) and birch (*Betula*) that are generally grown in temperate regions of Australia are described.

INTRODUCTION

Plant tissue culture is a method of rapidly cloning plants from isolated cells (e. g. microspores/ mesophyll cells), tissue (pith / cambium) or parts of organs (leaf segments/ nodal segments, root segments) under aseptic conditions in controlled condition of culture media, light, temperature and humidity. Tissue culture becomes an efficient and cost-effective method for rapid cloning of many woody ornamentals when cutting production is less efficient due to poor strike rates as observed with many tree species. Tissue culture is also a preferred method when the precious mother stock is very limited in supply, for example plants coming out of quarantine during the import process. Tissue culture can provide disease free stock of plants if the cultures are initiated from disease indexed mother plants free of contagious viral/ bacterial diseases. For example, the Quality Approved Banana Nursery (QBAN) program employing this method successfully supplies millions of Bunchy Top Virus (BBTV) free banana saplings across Australia. Disease free nature of carefully cultured plants also assists with overcoming quarantine barriers for international exchange of germplasm.

Tree species are less amenable to tissue culture cloning compared to herbaceous species. Temperate woody ornamentals like redbud, birch, maple and sycamore are species that are high in commercial demand but difficult to clone in large numbers. Although there are a few published papers on tissue culture of these species (Bowen-O'Connor et al. 2007; Cheong and Pooler, 2003; Girgžde, 2017; Huang et al. 2009; Wayne et al. 1995), the published protocols are not efficient for commercial cloning of specific commercial cultivars and varieties of these species. Therefore, I undertook this research to develop protocols suitable for commercial cloning of these plants.

MATERIALS AND METHODS

Published tissue culture formulations, MS Medium (Murashige and Skoog, 1962) and Woody Plant Medium (WPM) of Lloyd and McCown (1980) were modified with phytohormones and addenda to achieve efficient cloning of these species. Commercial supply of MS medium with vitamins and WPM medium with vitamins, plant hormones, laboratory grade sucrose and other addenda were sourced from Phytotech lab (https://phytotechlab.com), USA. Pure water prepared with RO system was used throughout the experiments. A factorial system trial with hormones and addenda was followed to determine suitable media combinations to achieve best results at stages 1-4 of tissue culture.

Culture media were sterilized at 121C at 1.2 kg/cm^2 pressure for 15 minutes before use. Sterilised media were stored at 22°C in the dark around 70% humidity for at least a week before use.

Disposable, commercial food grade containers (www.genfac.com.au) were used as culture vessels throughout. Cultures were incubated in an airconditioned clean room at 25±2°C, ≤70% relative humidity illuminated to 4000 lux with cool daylight fluorescent tubes.

Plants evaluated included three cultivars of redbud (*Cercis canadensis* 'Forest Pansy', 'Merlot' and Lavender Twist ['Covey']) and purpleleaf birch (Betula 'Royal Frost'). Young, 3-4 cm nodal explants from spring sprouts were used to initiate micropropagation of all the species studied. Detergent washed explants were disinfected for 5 min with 5% (v/v) commercial bleach (Clorox ®) followed by 5 rinses with sterile water before inoculating to initiation media.

RESULTS AND DISCUSSION

Culture initiation was better under dark incubation as browning of the explants followed by tissue death was significantly higher when incubated under light. Shoot growth and multiplication was invariably better under light incubation. To achieve faster multiplication the shoots were allowed to grow for up to 6 weeks, then used nodal segments from the axenic cultures for further multiplication. This method supplied 200-250 nodal explants per 500 ml culture vessel containing 10-12 multiplying cultures of redbud. Purple leaf birch was most proliferative and could generate 2,000 or more nodal segments per shoot cultures in a 500 ml jar in 6 weeks period. Purple leaf birch rooted on WPM medium supplemented with 0.1 mg/l IBA while redud cultivars rooted better ex vitro following incubation in MS medium supplemented with 2-5 mg /l IBA.

All the species were acclimatised in green house fitted with fine sprinklers to supply necessary levels of humidity. All the species acclimatised in a porous potting medium, when 90% or more relative humidity was maintained in the first week following the transfer. Gradual reduction in the humidity to 80% and 70% in the following weeks reduced fungal and bacterial infections on the plantlets and improved recovery of acclimatised plants.

CONCLUSION

Protocols for commercial cloning of a few redbud cultivars, purple leaf birch, maple and sycamore were developed. Micropropagation ability was higher in birch compared to redbud cultivars. Millions of in vitro rooted plantlets of these woody ornamentals per year can be generated at 10-20 US cents / plantlet using this protocol.

LITERATURE CITED

Bowen-O'Connor, C.A.., Hubstenberger, J., Killough, C., Marie, D., Leeuwen, V. and St. Hilaire, R. (2007). In vitro propagation of *Acer grandidentatum* Nutt. In Vitro Cellular and Dev. Biol. - Plant. *43*: 40-45.

Cheong, E. and Pooler, M.R. (2003). Micropropagation of Chinese redbud (*Cercis yunnanensis*) through axillary bud breaking and induction of adventitious shoots from leaf pieces. In Vitro Cellular and Dev. Biol. - Plant. *39*: 455-458.

Girgžde, E. (2017). Effect of cytokinins on shoot proliferation of silver birch (*Betula pendula*) in tissue culture. Environ.and Exper. Biol. *15*:1-5.

Huang, W.J., Ning, G.G., Liu, G.F. and Bao, M.Z. (2009). Determination of genetic stability of long-term micropropagated plantlets of *Platanus acerifolia* using ISSR markers. Biologia Plant. *53*:59–163.

Lloyd, G.B. and McCown. B.H. (1980). Commercially feasible micropropagation of mountain laurel (*Kalmia latifolia*) by use of shoot-tip culture. Comb. Proc. Intl. Plant Prop. Soc. *30*:421–437.

Murashige, T. and Skoog. F, (1962). A revised medium for rapid growth and bioassays with tobacco tissue cultures. Physiol. Plant.*15*:473–497.

Wayne, A., Mackay, W.A., Tipton, J.L. and Thompson, G.A. (1995). Micropropagation of Mexican redbud, *Cercis canadensis* var. Mexicana. Plant Cell, Tissue and Organ Cult. *43*: 295-299.

PROCEEDING'S PAPERS

EASTERN REGION OF

NORTH AMERICA

Dr. Charles W. Heuser, Jr., Regional editor

Seventieth Annual Meeting - 2021

Morton Arboretum – Chicago area, Lisle, Illinois USA

A Brief Look at Grafting *Franklinia* to ×*Gordolinia*

William Barnes

Barnes Horticulture, 2319 Evergreen Ave, Warrington, Pennsylvania 18976 U.S.A.

bill@barnhortservices.com

Keywords: ×*Gordolinia grandiflora*, *Franklinia alatamaha,* propagation, grafting

Summary

Franklinia as a landscape tree can have difficultires in container production as well as the landscape partly due to root system succeptibility to soil borne pathogens. ×*Gordolinia* is a hybrid of *Franklinia* and *Gordonia* appears to be more adaptable to varying soil types and could serve as a useful rootstock for *Franklinia*. This paper presents initial grafting method and post-grafting growth in *Franklinia* on a. ×*Gordolinia* rootstock.

INTRODUCTION

Franklinia ×*Gordolinia grandiflora* is a hybrid of *Franklinia alatamaha* and *Gordonia lisianthus* as performed and released by Dr. Tom Ranney and Dr. Paul Frantz (2006) Genetically the two species are quite close to one another. In some respects, ×*Gordolinia grandiflora* is superior to *Franklinia* because of its capability of adapting to varying soil types, a trait that *Franklinia* fails to share. Also, some container growers tell me that *Franklinia* is troublesome in production and often prone to soil borne fungal problems and sometimes the plants will look good one day and then fail the next. One grower in particular has had spectacular results with ×*Gordolinia* in containers with a high pine bark and sand media, so much so that a rooted 6-in cutting at the 1st of March is almost 4 ft high by the end of summer, when growing side by side to *Franklinia* the ×*Gordolinia* will be 2-3 times the size of the *Franklinia*.

If not sited correctly with that assessment sometimes a mystery, *Franklinia* cannot be grown in soils that ×*Gordolinia* will ordinarily thrive in, this opens the possibility that a *Franklinia* grafted to ×*Gordolinia* could function in soils that are not generally acceptable for *Franklinia*. It should be noted that while soil tolerance is true, ×*Gordolinia* is not as reliably cold hardy as *Franklinia*. This is due to ×*Gordolinia's* tendency to be non-photoperiodic and its close kinship to *Gordonia lisianthus*, generally recognized with limited hardiness.

Dr. Thomas Ranney in an email conversation to me found that grafting *Franklinia* to ×*Gordolinia* presents no special difficulties and the grafts proceed with vigor.

METHODS

Two-year-old ×*Gordolinia* growing in #2 pots with conventional bagged potting soil were selected as rootstocks. One was set aside as a control and *Franklinia* scions were selected in the fall of 2020, around late to early October from well-established plants at the Barnes Arboretum of Saint Joseph University (SJU), Philadelphia, Pennsylvania.

Franklinia scions were cut from a mature tree with good growth and flowering. Scions were 10-15 cm long with leaves removed. It was interesting that the pith in *Franklinia* stems is brown and not white. It is generally a good idea to avoid scions with significant pith in the center of the stem as pith can interfere with the grafting process. It was decided to proceed in spite of the presence of the pith. A traditional side graft (Figure 1) was performed with care being taken to minimize the surface area of the pith.

Figure 1. Completed active graft.

The base of the scion was wrapped with grafting rubber strips and then covered with an equal layer of parafilm grafting tape. After grafting the entire plant was enclosed in a white poly bag (Figure 2) and the bag was secured so that there was no air flow out of the bag in order to ensure a high humidity environment for the grafts. Leaves on the ×*Gordolinia* were kept and the plant was watered periodically to prevent dehydration. The completed grafted plant was kept in a shady portion of the greenhouse at the Barnes Arboretum greenhouse. Ambient temperatures were approximately 20-24°C during the day and about 15-17°C during the night. The grafted plant was left in the greenhouse for approximately 6 weeks and then removed to a cold greenhouse with the plastic bag intact for the rest of the winter.

Figure 2. Tent over completed graft.

Upon bud break in mid spring the grafted plant was moved outside and the plastic bag was removed. Care was taken to remove ×*Gordolinia* sprouts that would be in competition to the *Franklinia* shoots emerging. After the *Franklinia* shoots were allowed to grow for 6 weeks the grafting tape and the rubber bands were carefully removed and replaced with blue painter's tape to give additional support to the grafts. Both the grafted *Franklinia* plant and the comparison ×*Gordolinia* were keep in a standard nursery setting and fertilized while actively growing several times with liquid fertilizer at 250 ppm N. In late September grafts and control plant were compared.

RESULTS AND DISCUSSION

The control plant had numerous shoots with growth about 10-15 cm but no flowers. The grafted plant was kept clean of ×*Gordolinia* shoots and only the three *Franklinia* scions were allowed to grow (Figure 3). In late Sept the *Franklinia* showed vestiges of fall color and leaf senescence whereas the ×*Gordolinia* was growing actively and showed no such indicators of the approach of autumn. Photoperiodic vs no photoperiodic response? Presumably so but perhaps not a certainty as the effects of being grafted could play a part in such a response, at this point it is unclear which phenomenon is occurring.

Figure 3. *Franlinia alatamaha* grafts marked with blue tape on left, ×*Gordolinia grandiflora* on right.

The grafted plant had three shoots (Figure 3), one behaved as is typical of a central leader and the two grafts on either side put on some growth (15 cm) but not to the extent of the central graft which was close to (45 cm) which by all measures was surprising.

It seems that grafting of *Franklinia* to ×*Gordolinia* is successful as indicated by Dr. Thomas Ranney. Further work will be done to compare the growth rates of the ×*Gordolinia* and the grafted *Franklinia* on ×*Gordolinia* in typical soil conditions that is conducive to a host of other garden plants.

LITERATURE CITED

Ranney, T.G. and and Fantz, P.R. (2006). ×*Gordlinia grandiflora* (Theaceae): An intergeneric hybrid between *Franklinia alatamaha* and *Gordonia lasianthus*. Hortscience *41*:1386–1388.

Acknowledgements: Barnes Arboretum of the Saint Joseph University, Philadelphia, Pennsylvania. Supply of *Franklinia* scions and use of greenhouse space for allowing for the grafting secession to take place.

Tissue Culture Propagation of *Aronia melanocarpa*

William Barnes

Barnes Horticulture, 2319 Evergreen Ave, Warrington, Pennsylvania 18976 U.S.A.

bill@barnhortservices.com

Keywords: Micropropagation, black chokeberry, student lab

Summary

A micropropagation system suitable for a student lab is described using black chokeberry (*Aronia melanocarpa*). Explants formed five to six shoots in four to five weeks of culture. Rooted explants were moved to plastic boxes where they were hardened-off.

INTRODUCTION

The first experiments with plant tissue culture propagation occurred in 1902 but proved unsuccessful. By 1936 there was a renewed interest especially with the recognition and use of the plant growth hormones: auxins, cytokinins, and gibberellins. Experiments included taking established pieces of roots, thoroughly cleaning and placing in a nutrient broth of minerals, vitamins, sugars, mysterious entities from yeast extract, coconut milk and other sterilized plant derivatives. Murashige and Skoog made the first significant leap in mid 1960s. In the late 1960s new technologies were developed to grow plant tissues in aseptic culture.

Advantages of tissue culture

- Speed: Tissue culture is a quick process. In weeks, one can produce thousands of plantlets.
- Heath: Plants are disease free from being produced in a sterile environment.
- Flexibility: Plant growth can happen year-round, regardless of season.
- Space: Cultivators can grow ten times the plants in one-tenth the space of a regular grow operation.

- Innovation: Suspension culture opens the door for mutations and genetic engineering.

METHODS

The following content is based on plant propagation and physiology class, Spring 2021, At the Barnes Arboretum Horticulture Program. Instructor: H. William Barnes.

Plant tissue culture is defined as: The in vitro culture of plant protoplasts, cells, tissues or organs under controlled aseptic conditions which lead to cell multiplication or regeneration of organs or whole plants.

Other forms of asexual reproduction like, cutting, budding, and grafting are known as in vivo clonal propagation of plants. In vivo clonal propagation can be expensive, difficult, and unsuccessful. Plant tissue culture production or micropopagation is an alternative approach to in vivo production. The commonly used explants in micropropagation for initiation of the culture are meristem, shoot tip, and axillary buds. This tissue is multiplied in vitro (in glass).

Steps in the process include:

- The first step in micropropagation is the selection of stock or elite plants having desirable characters for their multiplication on a large scale.
- The next step in the process is to surface sterilize the tissue using various chemicals.
- After surface sterilization the explants are inoculated onto a medium and supplemented with various growth regulators, vitamins, and sucrose. *Aronia melanocarpa* (black chokeberry) was selected for the experiment and it was tissue cultured in a canning jar.
- The next step is the multiplication of explants. Each explant produces five to six shoots in a period of 4 to 5 weeks. Roots form with help from hormones and plants begin to form complete plantlets.
- Next the plants are hardened off. This involves the plants becoming resistant to stress, moisture, and disease. Plantlets must be protected from direct sunlight and humidity decreased. The plantlets develop roots during this period and cuticular wax is also formed in the aerial tissues.
- Finally, the plantlets become suitable for transfer to the field.

Aronia melanocarpa (black chokeberry) was selected for the experiment and it was tissue cultured in a canning jar. Roots form with help from hormones and plantlets begin to form complete plants. Each student got a small plastic container, tweezers and soil to fill the box (Figure 1).

Figure 1. Each students received a small plastic container, tweezers and soil to fill the box. Small, 3-inch-long plantlets were careful placed in soil using the tweezer.

Small, 3-inch-long plantlets were carefully placed in soil using the tweezer and pressed a bit to steady them. Each box was filled with 6 to 8 plant pieces (Figure 2).

The boxes were closed with an airtight lid as part of the hardening off process. The small boxes of plantlets were then placed in a large plastic container in the greenhouse to finish hardening off (Figure 3).

Figure 2. Each box was filed with 6 to 8 plant pieces. The boxes were closed with an airtight lid as part of the hardening off process.

Figure 3. The small boxes of plantlets were then placed in a large plastic container in the greenhouse to finish hardening off.

SOURCES

Barnes H. W. lecture April 04, 2021.

Bhatia, S., Sharma, K., Dahiya, R., and Bera,T. (2015). Modern applications of plant biotechnology in pharmaceutical sciences. ProQuest Ebook.

Gibberellin and Clipping Promote Germination in Fresh Grape Seeds

Bridget Bolt, Roberta Magnani, Marta Nosarzewski, Carlos Rodríguez López and Robert Geneve

Department of Horticulture, University of Kentucky, Lexington, Kentucky 40546 U.S.A.

Rgeneve@uky.edu

Keywords: *Vitis*, dormancy, physiological dormancy, stratification

Summary

An initial experiment was conducted to reduce or bypass the stratification requirement for dormancy release and germination in grape seed. By utilizing fresh seed from mature fruit that had not completed the final maturation drying stage of development was found to be induced to germinate after a 2000 ppm gibberellic acid treatment or after clipping the distal end of the seed. This effect was further enhanced by combining the gibberellin and clipping treatments yielding 100% germination in this preliminary study.

INTRODUCTION

Grape is commercially important both as a table fruit, a processed fruit for raisins, juice, and jams, as well as for wine production. Selected cultivars are commercially propagated by hardwood cuttings or grafted on resistant rootstocks (Davies et al., 2018). However, weather patterns around the world are changing and there is a need to breed and propagate new adapted selections of a variety of traditional crops including grape. For grape, there are predictions for dramatic reductions (up to 81% by the late 21st century) (White et al., 2006) of suitable wine grape acreage in the United States. Seed germination is an important step in traditional breeding programs as well as those "accelerated crop breeding" programs utilizing novel genetic approaches.

Grape seed has physiological dormancy and requires three to four months of chilling stratification. A system that could bypass this stratification time to expedite seedling production could reduce breeding cycles and facilitate novel "accelerated breeding" programs. Therefore, the objective of this project was to investigate the impact of partial seed coat removal and gibberellin treatment on germination of freshly harvested grape seeds.

MATERIALS AND METHODS

Seeds were extracted from ripened grape (*Vitis* 'Cabernet Sauvignon') fruits from greenhouse grown plants. Fruits were physically crushed by hand and the pulp removed from seeds by rubbing with paper towels. Seeds were surface disinfested for 10 minutes in a 10% commercial bleach solution followed by three rinses in sterile distilled water. Half the seeds were left intact, and half were cut through the seed removing the distal rounded portion of the seed. Intact and cut seeds received a 24-hour soak in sterile distilled water or a 2,000 ppm filtered sterilized gibberellic acid (GA) solution. Seeds were then placed in Petri dishes on an in vitro Bacto-agar based Murashige and Skoog salts medium without sucrose (Figure 1). Germination was at 25°C with 16-hr light.

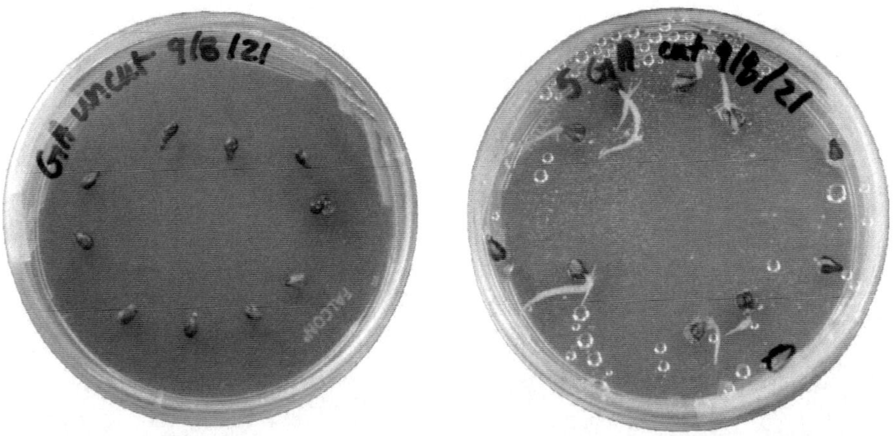

Figure 1. Seeds were sown on MS in vitro medium without sucrose.

RESULTS AND DISCUSSION

Untreated seeds failed to germinate (Figures 1 and 2). All seeds that were cut and treated with GA germinated after about 14 days. Intact seeds treated with GA or cut without GA germinated at 40% or 20 % respectively, but germination was slow taking between 20 and 40 days to initiate germination. Seeds germinated in vitro transitioned to produce seedlings, but further observation is necessary to ascertain the vigor of these seedlings (Figure 3).

There are several surgical methods to by-pass seed physiological dormancy including embryo removal from the seed or disrupting seed coat integrity (Geneve, 1991). Hormones, primarily GA, can also substitute for chilling stratification to satisfy dormancy (Baskin and Baskin, 2014). There is also significant anecdotal evidence that utilizing fresh seed that has not gone through the desiccation process can show less dormancy compared to dried seeds in several woody perennials including alder (*Alnus*), persimmon (*Diospyros*) and eucalyptus (Schopmeyer, 1974).

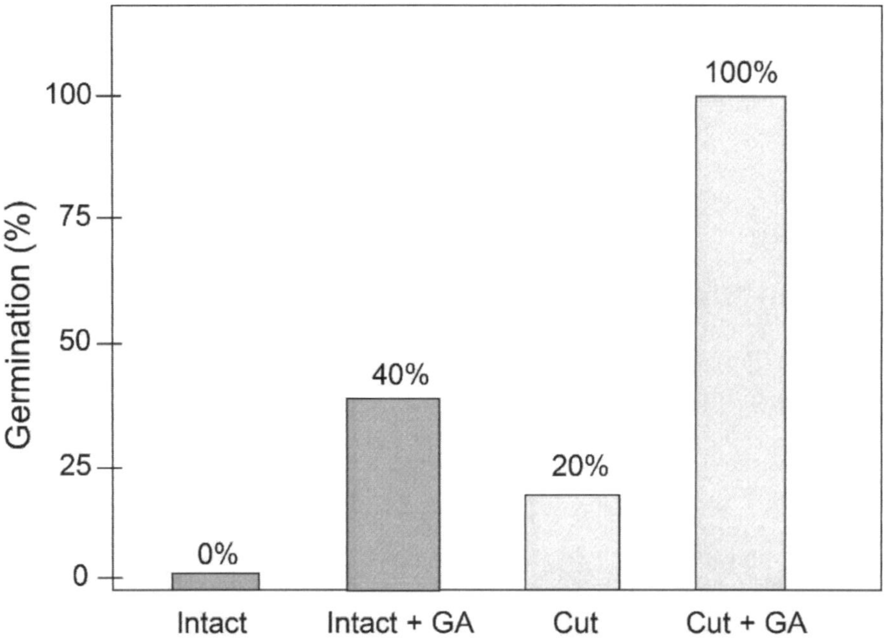

Figure 2. Germination in fresh grape seeds after being cut and treated with 2,000 ppm gibberellic acid.

Figure 3. Seedling from seed that was cut and treated with 2,000 ppm gibberellic acid.

The present preliminary study with grape demonstrated that seeds that have been cut and treated with GA could germinate and transition to seedlings thus reducing the time to produce a seedling compared to traditional stratification treatments. This study was done with fresh seed and additional studies are underway to see if the combination of cutting and GA treatment is only efficacious in fresh seeds or can also be applied to dried and stored seeds.

LITERATURE CITED

Baskin, C. C., and Baskin, J. M. (2014). *Seeds: Ecology, biogeography, and evolution of dormancy and germination*, 2nd ed New York: Academic Press.

Davies, F. T. Jr., R. L. Geneve and Wilson, S. B. (2018). *Hartmann and Kester's Plant Propagation: Principles and Practices.* Boston: Prentice-Hall. Ninth edition.

Geneve, R.L. (1991). Seed dormancy in eastern redbud (*Cercis canadensis* L.). J. Amer. Soc. Hort. Sci. *116*:85-88.

Schopmeyer, C. S., ed. (1974). Seeds of woody plants in the United States. *U.S. For. Ser. Agr. Handbook 450.*Washington, DC: U.S. Govt. Printing Office.

White, M. A., Diffenbaugh, N. S., Jones, G. V., Pal, and J. S., Giorgi, F. (2006). Extreme heat reduces and shifts United States premium wine production in the 21st century. Proceedings of the National Academy of Sciences, *103*:11217-11222.

Tech Talk: Bane or Boon? A Brief Look at Horticultural Tech

Lincoln Gillon

Senior Team Leader, Propagation Department, Prides Corner Farms, Lebanon, Connecticut 06249 U.S.A.

lincoln.gillonpcf@gmail.com

Keywords: New challenges, information, research and development, collecting data, sensor equipment, environmental factors

Summary

Current technology allows us to grow quality plants, while reducing many of the input costs - like time and labor. In choosing new technology, be selective. It is not necessary to adopt every new gadget, app, or service that comes down the pipeline. Much research and development are out there waiting to help us with this. R&D, Use the information out there. But you should be aware of many growing factors, and utilize tools to tweak things one way or another.

INTRODUCTION

Prides is a wholesale nursery in Eastern Connecticut with about 500 acres under production across four locations.

I have always loved and enjoyed plants, whether that was my 300 houseplants while growing up, or at my first job-riding my bike after school to work a couple hours at a local greenhouse, doing some landscape maintenance over the years, or floral design along with my wife Lisa in her wedding flower business. Currently I manage our crew in Harvesting field cuttings, and along with my colleagues, direct the sticking, rooting, and grow-on of our internal liner production. We grow fruits, vegetables, trees, perennials and shrubs. With millions of liners to produce we have to learn, we have to grow, and we need to produce results. This talk is designed to raise questions, pose a few answers and inspire research and investigation.

There have been many changes in the past century of horticulture. All of these changes came with challenges, seen or unforeseen, that push us to grow, to learn and to understand more aspects of what makes a healthy plant. A happy plant. In many respects as horticulturists moved further from nature we have had to get better at nurture. Spoiler alert: There is more to growing than sunlight and H_2O.

Current technology allows us to grow quality plants, while reducing many of the input costs - like time and labor. In choosing new technology, be selective. It is not necessary to adopt every new gadget, app, or service that comes down the pipeline.

Some changes in industry practices, which are designed to maximize space, reduce residency time and streamline transport to customers, have affected many growing factors. Container and tray sizes impact many soil aspects, including changes in volume, weight, composition, moisture, temperature and root-to-shoot ratio.

Every advance leads to new challenges. A simple advance — moving from ground to pot or container - yields the question: What do you put in the pot? Dirt? Soil? Media? Where is the technology? It is in things we can measure such as weight, porosity, pH, EC, soil nutrients. These are all things we can measure. But what is our target? Education and Research can help us find an Optimum range. Simply the ability to take samples and test pH or EC on our own without sending out to a lab increases our ability to react and make adjustments in our grow program. This is especially important if lead times continue to shrink, tightening of sales windows or changing customer needs.

Much research and development are out there waiting to help us with this. R&D, Use the information out there. In today's world going it alone does not make much progress. Much like mathematics we need to build and grow on prior work and knowledge. Why mention this you ask? I want to emphasize the reasons we are gathered in person and online; why we belong to IPPS; To seek and to share. A person could pick one genus or even one species and study it for a lifetime. One could also spend a lifetime reading innumerable books and internet articles (preferably those ending in --- edu which are more reliable sources of information than simply "Bob's Garden blog"). Another tool that has become more available in the past year is collaborating with colleagues over video conferences.

But where is the tech with a capital "T"? It is in the data. Technology available today can gather infinite data points. We have moved beyond daily high/low temperature records. Sensors can collect information all day and all night. Pair them with software and that data can be collated and graphed. It can be turned from numbers into pictures. And as we all know "a picture is worth a thousand words". I for one have found a graph to be nothing more than a pictorial representation of data. Easier to see, easier to understand the current conditions, where they are going, and most important where they have been. We cannot collect data for data's sake alone. Data is only useful if accurate and one is able to understand the correlation to the factors we are trying to measure or control. Technology sitting on a shelf, or a dusty corner is never effective. Whether that technology is simply a handheld pH or EC meter, high-tech control system, pruning machine, mulching machine, or fertigation injector.

You've heard that those who don't learn from history are doomed to repeat it. When plants yield results, whether good or bad, without reliable information; you are not doomed to repeat those results. You cannot know what factors went right or wrong and therefore repeat outcomes or avoid pitfalls. This is where collecting data pays dividends. A grower quote "Don't fly blind, use your diagnosis and records as a tool for now and into the future." I will grant that an experienced grower can continue to grow year after year relying on personal knowledge. But that breadth of knowledge has become harder to source and hire, as companies get bigger, competition changes the playing field, and new talent (as in many industries) finds specialization more prevalent.

Mist timers and clocks are simple and effective tools to maintain a desirable rooting environment in propagation. Current models allow for programmable changes to mist frequency during the day. A simple example of a misting day could be to have three programs. 'Program A' running from 8 AM to 11 AM at 30-minute intervals; "Program B' running 11 AM - 3 PM every 20 minutes increasing the mist frequency during the heat of the day, and 'Program C' from 3 PM to 8 PM misting every 25 minutes. The ability to program this in advance can eliminate some of the labor involved in making manual changes throughout the day. But time alone is not the only factor. When the weather changes or crops age, growers benefit from making manual adjustments to maintain the growing environment. The newer mist clocks available on the market have added functionality, such as light sensors (to reduce mist during cloudy periods) which can make some automatic adjustments.

Irrigation valves can be automated with timers, but share similar limitations. Electronic and wireless options are now on the market. These provide more flexibility for remote access to modify your watering schedules. When conditions change, remote accessibility allows a grower to make necessary adjustments without having to be on the site of every irrigation valve. Manually changing valves uses a lot of labor, which can be significantly reduced by electronic valves. There are even some wireless valves available with bluetooth capability. Wireless models will reduce the complexity and cost of installation as well as maintenance on the system, as wired units are subject to damage from rodents and equipment. In my personal experience, I have experienced a change from manual ball valves to wired irrigation clocks. This saved more than an hour a day in manually changing more than a dozen valves. A grower will still need to oversee systems to make sure the equipment runs properly. More than once when cycles failed to run, I found evidence of cleanly cut wires which appeared as sabotage, but upon closer inspection, the culprit was identified by the pile of droppings Peter rabbit left behind. The takeaway is: utilize the technology available, but be conscious of its limitations and ensure there are failsafes in place to protect your crops.

Another example is advancing from individual thermostats to control heating, ventilation and airflow to dual or multiple stage thermostats which enable these environmental factors to be managed without competing against each other. When these are on individual thermostats, overlap can occur causing heating and cooling mechanicals to operate simultaneously causing a bad situation. A common greenhouse heater is an oil-fired furnace, which can backdraft and cause a fire if the ventilation system reverses the exhaust airflow, or burn-out heating equipment, requiring expensive repairs (sometimes at 3 AM. in sub-freezing temperatures). In a less extreme scenario, it can simply burn fuel excessively, hurting the bottom line, especially if it goes undetected. Additionally, depending on the crops, exhaust gases from continuously running equipment can adversely affect plant health. These situations can occur due to placement of the individual thermostats, aging equipment, or human error. The newer multistage thermostats provide lockout control maintaining separation of heating, cooling, and ventilation functions. With a single multistage unit, placement and wiring are simplified. The possibility of multiple units being accidentally adjusted is also eliminated, reducing the human error factor. Oversight is still required to manage set-points and make adjustments for different growing conditions and the weather at your location.

Another important tool at your disposal is the sensor equipment to help trigger or monitor these various systems. Sensors are available to monitor the temperature of soil and air (in the greenhouse and ambient). Soil sensors can measure pH, EC and moisture. Other sensors for those critically important light levels can measure PAR and DLI (daylight integral). You can monitor humidity in the greenhouse, vapor pressure deficit (VPD) and other factors in your facility. Without sensor data, you might be growing by the seat of your pants, which may work for some growers with many years of trial and error under their belts. Before sensors were available to quantify these various growing factors, human input was required to make decisions and adjustments on an ongoing basis. Labor and time were required to take soil samples and ship them to the lab for pH and EC analysis. Temperature and humidity were read manually in each growhouse, which would then require adjustment. Every adjustment required the judgement of a knowledgeable grower, who needed to be onsite to make those adjustments, often multiple times a day, within each growing area. This becomes completely untenable in a larger operation. When growing conditions are not monitored and adjusted in a timely fashion, crops are subject to additional stress, leading to plant health issues or, ultimately, crop losses. When data is easily accessible, growers can make more informed, proactive adjustments, instead running around like headless chickens, doing damage control.

Although wired sensors are available, it's not practical to have multiple wired systems in each grow area monitoring all the necessary data. Wired sensors can also be prone to damage by pruning equipment, which I have witnessed. There are now wireless sensors on the market, which are simple to install and use, without the hassle of running wires to half a dozen sensors. I have installed dozens of them myself; each in a minute or so. The Bluetooth capabilities of many of these sensors allow growers to monitor conditions, receive alerts and even set alarmed parameters, while walking through the grow area. By using a WiFi gateway, the visibility of these sensors can extend to the internet, and be available on any device with internet access. To the grower at home, or the on-call nursery crew, this means fluctuations can be identified immediately, minimizing potential crop damage. Case in point, I received an alert one night when one of our greenhouses fell below temperature. I was able to pull up the sensor data that indicated a heater failure. Our on-call crew was dispatched, and the issue resolved before any crop damage could occur. Having sensors to monitor and track data on an ongoing basis is much more cost effective and reliable than periodic, labor-intensive manual data-collection. Moreover, a single adverse event that is prevented by sensor data will easily offset the upfront equipment costs.

Taking the next step up with a control system. These can group even more factors. A control system can manage environmental factors within a growhouse, react to external weather, and help with irrigation, fertigation, and misting. Tying everything together in sync like never before. This of course comes with new challenges and requires adjustments in thinking as we balance the various growing environment targets using prior knowledge and adding new sensors for a greater understanding of the intricacies that affect various crops. Never fear! You can still fly by the seat of your pants, but do so while gathering multiple data points to quantify and graph your environment. Sometimes mother nature gives us perfect conditions for rooting or growing, but more often not. The more tools we have to supplement what nature supplies, the more consistent our results can be.

The good news: don't be overwhelmed; you don't have to know it all — I certainly don't. But you should be aware of many growing factors, and utilize tools to tweak things one way or another. New technology is not a necessity, but it can help us build on all the knowledge base we have at our disposal. Change cannot be forced; it simply arrives sooner or later. Our decisions determine whether we're in the wave, at the forefront, riding the coat-tails, or left paddling in the wake. Run your CBAs and ROIs (cost benefit analysis and return on investment) but do so while painting the whole picture. It's cheap and easy to get a few sensors or test probes. The power to drive great change might fit in the palm of your hand.

Chicago Botanic Garden Plant Evaluation Program

Richard Hawke

Chicago Botanic Garden, 1000 Lake Cook Road, Glencoe, Illinois, 60022 U.S.A.

rhawke@chicagobotanic.org

Keywords: Plant evaluation program, perennial plants, botanical garden, weed risk assessment, selection criteria

Summary

The Chicago Botanic Garden's Plant Evaluation Program was established in 1982 and is currently one of the largest and most diverse evaluation programs. The program focusses mainly on herbaceous plants, but some woody plants are included. For comparative trials, commercially available species and cultivars within specific plant genera and grow in side-by-side for easy comparison of traits and performance. Invasive plant trials undertaken include numerous taxa from international collecting trips as well as common garden plants.

INTRODUCTION

The Chicago Botanic Garden's Plant Evaluation Program was established in 1982 and is currently one of the largest and most diverse evaluation programs focusing primarily on perennial plants in the U.S. There are six components of the program: comparative trials, cooperative trials, green roof, plant exploration, invasive plants, and special projects.

Comparative Trials

For comparative trials, we acquire commercially available species and cultivars within specific plant genera and grow them side-by-side for easy comparison of traits and performance. There are several criteria we consider when selecting comparative trials, including:

- Important horticultural or garden genera, such as salvia, geranium, and phlox,
- Genera with significant breeding developments, such as *Echinacea and Coreopsis*,
- Genera or plants that are at the time uncommon locally, such as betonies, geums, and potentillas,
- Winter hardiness/adaptability, such as *Stokesia*, English shrub roses, and gentians,
- Cultural adaptability to conditions of site, such as soils, moisture, and exposure,
- Specific disease issues, such as powdery mildew and rust.

There are 18 targeted trials underway in 2021, including *Anemone, Astrantia, Baptisia, Buddleja, Calamagrostis, Calycanthus, Chrysanthemum, Deschampsia, Echinacea, Hibiscus, Kniphofia, Leucanthemum, ×Mangave, Phlox, Physocarpus, Salvia, Sanguisorba*, and *Silphium*.

Our trial gardens are open to the public. The primary trial garden (Figure 1, left) is a full-sun site with 7.5 pH, well-drained to periodically moisture retentive clay loam soils, and no wind protection. A new shade trial garden (Figure 1, right) is opening in October 2021; the initial comparative trials include *Abelia*, *Ajuga*, *Bergenia*, *Carex*, *Clematis*, *Helleborus*, *Hydrangea arborescens*, and *Pulmonaria*.

Figure 1. Primary trial garden (left) and new shade trial garden (right).

The primary advantages of comparative trials are:

1. Compare old and new cultivars	2. Evaluate uncommon or underused garden plants	3. Evaluate new and unreleased plants from the Garden's breeding program and other plant breeders

Figure 2. Comparative trial examples (left to right) coneflower, burnet, *Amsonia*.

There are four broad evaluation criteria:

1) Cultural adaptability to the cultural and environmental conditions of the site,
2) Winter hardiness and adaptability,
3) Disease and pest resistance,
4) Ornamental traits related flowers, foliage, and habits.

Cooperative Evaluation Trials

We have longstanding relationships with a variety of commercial cooperators including:

Plant introduction programs and nurseries

- Donated plants,
- Source of new and older plants—often the starting point for a comparative trial,
- Plants are typically introduced but often not yet readily available,
- Do not always have control of everything that is sent for trial, which results in plants that do not fit into a comparative trial,
- Annual evaluation results are reported to cooperators.

All-America Selections perennials trial

- New national program began in 2016,
- Plants are evaluated in a three-year trial cycle,
- Annual evaluations contribute to AAS winner status.

Plant breeders

- Less common these days due to breeders/companies doing their own trials,
- Has always been a challenge because our gardens are open to the public.

Botanical gardens and arboreta

USDA—NC-7 Regional Ornamental Plant Trials—woody plants only

Plant societies, such as American Boxwood Society and Holly Society of America

Green Roof Trials

The Garden has been evaluating herbaceous and woody plants for green roof culture since 2010, utilizing two distinct green roofs on our Plant Conservation Science Center (Figure 3).

The goal of the green roof trials is to add taxa to a national compendium of plants grown on extensive to semi-intensive green roofs.

The goal of the green roof trials is to add taxa to a national compendium of plants. One 8,000 square foot green roof features North American native plants at the species level and with minimal maintenance provided. The result is a wilder or more naturalistic landscape.

The other 8,000 square foot green roof grows native and exotic plants including cultivars. This roof has increased maintenance to keep plants in place for a more garden-like display. The green roofs are used by graduate students working on various projects related to pollinators, plant communities, ant populations, and "soil" microbes.

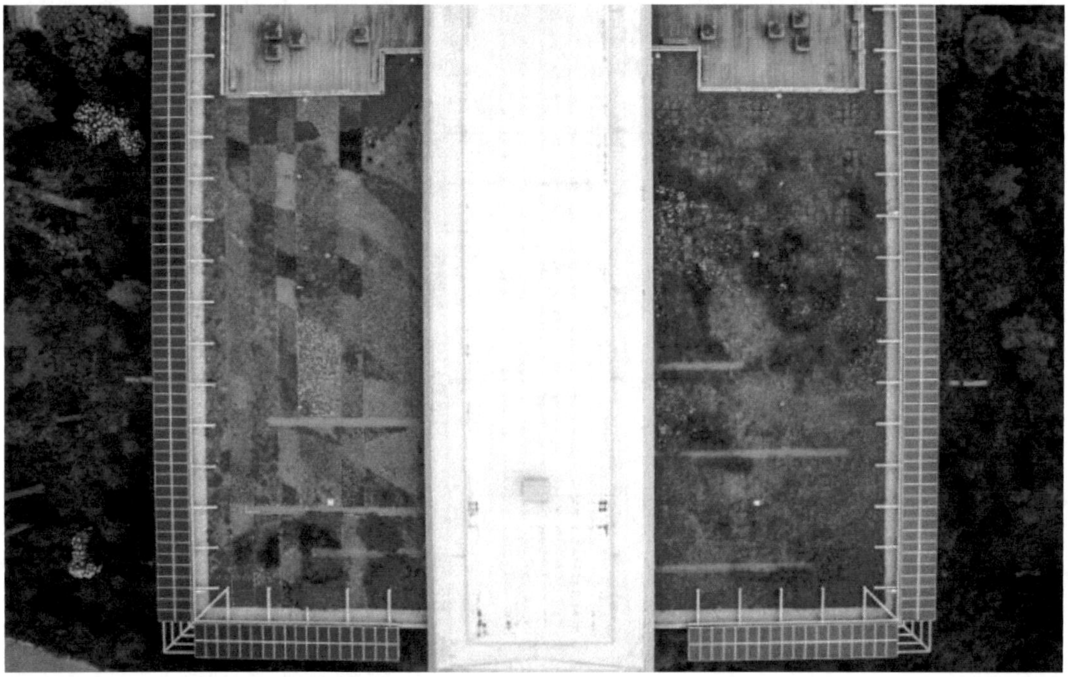

Figure 3. Two distinct green roofs on our Plant Conservation Science Center (Figure 3).

Plant Exploration Program Trials

The Chicago Botanic Garden in collaboration with other botanical institutions has participated in national and international collecting projects. Plant exploration programs trials include: South Korea, China, Altai Republic/Central Siberia, Russian Far East, Republic of Georgia, Uzbekistan, United States (Ozarks, Southeast, and Black Hills).

We target geographic regions with plants that are adaptable to our current and future climatic and growing conditions. Plants from international expeditions undergo a weed risk assessment process; taxa that don't get a pass or fail determination must be trialed.

Invasive Plant Trials

Invasive plant trials undertaken include numerous taxa from international collecting trips as well as common garden plants. A weedy designation has always been part of the evaluation criteria but considering the potential invasiveness of any species is essential. Examples of invasive trials we have completed are *Buddleja* for reseeding potential; *Miscanthus* for invasive potential based on seed viability; *Persicaria* (*Fallopia* and *Polygonum*); and wild-collected exotic taxa that dropped out of the weed risk assessment process.

Special Evaluation Projects

Periodically, we initiate projects that are independent/outside the scope of the comparative or cooperative trials. Examples of recent and upcoming special projects include 1) tender perennials used as annuals; 2) comparative trial of various umbellifers (not a genus trial); and 3) nativar project that observes pollinators on specific native species and related cultivars and hybrids. Nativar projects include black-eyed Susan (Figure 4), wild geranium, smooth penstemon, New England aster, and aromatic aster. Project participants include Chicago Botanic Garden, Denver Botanic Garden, Mt. Cuba Center, and San Diego Botanic Garden.

Figure 4. Example of a Nativar project looking at black-eyed Susan.

Results and Reporting

Reporting evaluation results through publication and outreach is important throughout the term of a trial (Download at www.chicagobotanic.org). Our primary publication is *Plant Evaluation Notes*, which is a periodic publication of Chicago Botanic Garden, and focuses on completed trials.

A long-running feature in *Fine Gardening* reports on our trials, both on-going and completed. Additional reporting is published in various garden and industry journals and websites, and through outreach via lectures, tours, and classes (Figure 5).

Figure 5. Two examples of reporting evaluation results through publication and outreach.

Breeding New Plants and a New Breeder

Brent Horvath

Intrinsic Perennial Gardens, Inc., 10702 Seaman Road, Hebron, Illinois 60034-9535 U.S.A.

brenth@intrinsicperennialgardens.com

Keywords: Plant breeding, herbaceous perennials, propagation, mentoring

Summary

The breeding program at Intrinsic Perennial Gardens is discussed with special emphasis on the criteria for breeding and selecting new cultivars. Successfully breeding new plants comes from years of observation including taking inspiration from peer mentors. It is also important to mentor the next generation of breeders.

INTRODUCTION

What is plant breeding and how does it differ from other methods for obtaining new plant introductions such as selecting?

Plant breeding is more intentional, can be done at home or at work, and is preformed with specific goals in mind. In contrast, plant selection can be more haphazard, made in or by nature, and is opportunistic.

Reasons for plant breeding

There are many reasons to breed new plants. Included would be the following:

- Hybrid vigor
- New colors
- Height differences, usually shorter
- Disease resistance
- Sterility or a longer bloom time
- Rebloom
- Foliage that differs from the parent, selection based on variegated and colored foliage

Getting a new plant to market includes a number of steps including:

- Naming the new plant
- Propagate the new plant
- Decide who to give or sell it to
- Decide to protect it or not

Propagating at Intrinsic

Plants are the nursery are primarily propagated from cuttings or seed (Table 1).

Table 1. Common plants at Intrinsic nursery and their primary propagation method.

Cutting produced plants	Top Seed produced plants
Amsonia	*Calamagrostis brachytricha*
Aster taxa	*Echinacea*
Geum	*Heuchera*
Leucanthemum	*Liatris*
Nepeta	*Penstemon digitalis* 'Husker's Red'
Rudbeckia	*Rudbeckia fulgida* var. deamii
Sedum	*Schizachyrium scoparium*
Stachys taxa	*Sporobolus heterolepis* and other native grasses
Bought as unrooted cuttings	Division
Aster	Allium
Monarda	*Andropogon* also doing some from tissue culture
Salvia	*Bouteloua gracilis* 'Honeycomb'
Veronica	*Calamagrostis*
	Carex- many species
Tissue Culture	*Festuca* 'Cool as Ice'
Andropogon	*Molinia*
Geranium	*Panicum*
Geum	*Pennisetum*
Polemonium	*Schizachyrium*
Rudbeckia	*Sesleria*
Seed collecting and cleaning is a necessity for a breeder. However, we also buy seed Prairie Moon Nursery and Jellito Perennial Seeds.	

Identifying potential new plants

How do you learn a new plant?

- See it in catalogs or pictures or advertisement
- See it in real life
- Have someone recommend it
- Read about it in a book or magazine
- But you really only learn a new plant by planting it and observing it!

Mentoring

My inspiration and my mentors:

- My high school horticulture teacher — Jeff Yordy
- My Dad
- Alan Bloom — specifically his book Hardy Perennials
- George Radtke — I can always call George
- Roy Klehm
- Roy Diblik — teaches me something every time we meet.

As a personal mentor, I have been working with a new breeder — Jake Letmanski (Figure 1); I got lucky, but I hope this inspires you to mentor someone in your company.

Figure 1. Jake Letmanski, Arie Blom, and Brent Horvath (left to right).

Jake Letmanski came to me at 15 years old as a hobby breeder and I quickly turned him into a commercial breeder (Figure 2).

His grandparents are farmers and his mom is a school teacher. He has a wide interest in different plants. So, I fed him with a wide range of new plants and shared a lot of information and books with him.

I helped him on his search for a college with horticulture by reaching out to fellow breeder Rick Grazzini. I've encouraged and helped his pet projects including desert willow, and other woodies including *Ceanothus*.

I've introduced him to other plant people and breeders. Including Rick Grazzini, Mike Yanny and Kim Shearer who intern introduced him to Jeffery Carsten with USDA.

Today he is a sophomore at Iowa State studying agronomy and horticulture. What does success in breeding look like to him: peer, trade and consumer acceptability. In 3–5 years he hopes that people know more about the benefits of garden plants besides just the ornamental qualities.

His favorite books are *The Manual of the Trees of North America* by Charles Sprague Sargent and *Jewels of the Plains* by Claude Barr. His favorite plants include peonies, *Silphium laciniatum* (the compass plant), and oak trees.

His obstacles are the learning curves related to a new genera and new people.

Figure 2. *Dianthus* 'Fuchsia Fire' pp#14,895.

Brown Turkey Fig Softwood Cutting Propagation

Allie Maternowski[1], Winston Dunwell[2], Dwight Wolfe[2], Virginia Travis[2] and Daniel Becker[2]

[1]Department of Horticulture, Murray State University, 201 South Oakley, Murray, Kentucky 42071 U.S.A., [2]Department of Horticulture, University of Kentucky, Lexington, Kentucky 40546 U.S.A.

winston.dunwell@uky.edu

Keywords: *Ficus carica*, cutting propagation, softwood cuttings, rooting hormone, mist propagation

Summary

A shortage of hardy common fig cultivars in 2020 led to a study to produce plants of brown turkey fig that could be grown in containers for fruit production. Softwood cuttings of *Ficus carica* 'Brown Turkey' were collected in June 2020, scored and dipped in Hormodin 2 talc, and then placed in 50% : 50% by volume perlite : pine Bark and Pro-Mix BX : pine bark substrates. These cuttings were evaluated for rooting a month later. In 2021, this method was repeated but with 100 cuttings. The cuttings were collected in June 2021 from the fig plants that grew from cuttings in 2020, and they were evaluated for rooting a month later.

INTRODUCTION

In May of 2020 nurseries that grow cultivars of *Ficus* spp., known for hardiness and a parthenocarpic (fall) crop that would ensure a harvest each season, were out of stock. In order to start a trial growing figs in containers for plant production practices and to be transplanted into 25-gal containers for fruit production, we would have to produce the liners needed. Considering the shortage, a trial on softwood cutting propagation was initiated.

Ficus carica 'Brown Turkey' plants were needed for a container grown fig fruit production research project. Cutting wood was provided by a colleague. No literature was found on soft wood cutting propagation of figs. Personal communication, indicated figs rooted easily from softwood cuttings. Most figs are propagated by hardwood cuttings that are callused and placed directly into the ground or rooted in containers. (Hartman and Kester, 2011).

MATERIALS AND METHOD

On June 5, 2020, 40 cuttings were divided into two groups. All cuttings were single wounded 0.5 inches (1.27 cm) from the basal end and dipped in Hormodin 2 IBA Talc [3000 (0.3%) ppm]. One group was placed in perlite : pine bark at 50% : 50% by volume and the other in Pro-Mix BX : pine bark at 50% : 50% by volume. Cuttings were placed in community trays. The cuttings were further subdivided into those considered vigorous (Tray – 1) and those determined to be of lesser quality (Tray – 2).

All were placed in a mist propagation bed with 10 seconds of mist every 10 minutes. Those of lesser quality would be considered bonus plants if they rooted. All cuttings were rated on 0 to 5 scale with 0 indicating no rooting and 5 indicating the best rooting (Figure 1). In order to be able to replicate future cutting production of 'Brown Turkey', records were maintained.

Figure 1. Rooting rating scale 2020; left to right 0, 1, 2, 3, 4, 5.

The 2020 study was replicated in 2021 with 100 cuttings beginning on June 3, 2021 following the same propagation practices. There were five trays of cuttings in each treatment placed in perlite: pine bark, and Pro-Mix BX : pine bark, each at 50% : 50% by volume. All trays were placed in a mist propagation bed with 10 seconds of mist every 10 minutes (Figure 2). After noticing sun damage, a shade cloth was placed overhead with 50% shade. The cuttings were rated on the same scale as before (0 to 5 scale) (Figure 3).

Figure 2. Trays in mist propagation bed 2021.

Figure 3. Rooting rating scale 2021; left to right 5, 4, 3, 2, 1, 0

RESULTS AND DISCUSSION

On July 8, 2020, the cuttings were evaluated (Table 1). The vigorous cuttings all rooted. Of the lesser quality cuttings, 4 did not root (0) and 4 rated 1 on a 1-5 scale. Cuttings in Trays 1 are shown in Figure 4.

Table 1: Rooting rating based on substrate and cutting quality.

Rating	Perlite : Pine Bark		Pro-Mix BX : Pine Bark	
	Tray 1	Tray 2	Tray 1	Tray 2
0		3		1
1		3		1
2	5	2	4	2
3	2		4	3
4	2	1	2	2
5	1	1		1
Average	2.9 a*	1.6 b	2.8 a	2.7 ab

*Means followed by the same letter are not statistically different (least significant difference) (P>0.05).

The images and observation rating indicate that the substrate did not influence rooting. Looking at Figures 4 and 5 and rating observationally based on white roots there appears to be a Pro-Mix BX : pine bark advantage. The quality of the cuttings influenced rooting of softwood fig cuttings in the evaluation. All the rooted cuttings grew vigorously. Three plants did not overwinter in an unheated nursery quonset with a single layer of white poly cover.

Figure 4. Rooting of Pro-Mix BX : pine bark 2020.

Figure 5. Rooting of Pro-Mix BX : pine bark 2021.

On July 16, 2021, the cuttings in the repeated study were evaluated (Table 2). The cuttings placed in the Pro-Mix BX : pine bark substrate had a higher average rating than the cuttings placed in the perlite : pine bark substrate. Because the Pro-Mix BX : pine bark substrate has a higher water-holding capacity, the cuttings were less affected by sun damage in the beginning stages of rooting.

The images and statistics indicate that the substrates did not have an influence on rooting because there is not a large enough difference between the two substrates; however, there is a slight Pro-Mix BX/Pine Bark advantage considering the higher observed average rating and percent of cuttings with roots.

Table 2: Effect of Substrate on rooting and quality of roots of fig cuttings.

Substrate	Percent with roots (survival)	Average rating
Pro-Mix BX : pine bark	92	3.84
Perlite : pine bark	84	3.50
Mean	88	3.67
LSD[1] (0.05)	16.2	0.71

[1]Least significant difference – means that are less than the LSD are not statistically different at the 0.05 probability level.

LITERATURE CITED

Davies, F. T. Jr., R. L. Geneve and Wilson, S. B. (2018). *Hartmann and Kester's Plant Propagation: Principles and Practices.* Boston: Prentice-Hall. Ninth edition.

Supplemental Nickel Corrects Mouse Ear Disorder of Bitternut Hickory

Brandon Miller[1] and Nina Bassuk[2]

[1]Department of Horticultural Science, University of Minnesota, 1970 Folwell Avenue, St. Paul, Minnesota 55108 U.S.A. [2]School of Integrative Plant Science, Cornell University, Ithaca, New York 14853 U.S.A.

bmmiller@umn.edu

Keywords: *Carya cordiformis*, mouse ear disorder, nickel deficiency

Summary

Mouse ear disorder manifests as leaf curling, necrotic margins, rosetting of the stem, suspended leaf expansion, and stem dieback in certain woody taxa including river birch and pecan. This is the first report of mouse ear disorder on bitternut hickory (*Carya cordiformis*) and its correction in plants treated with supplementing nickel either as a substrate drench or foliar application.

INTRODUCTION

In the early 2000s, it was discovered that mouse ear disorder, an issue limiting the production of river birch (*Betula nigra*) and pecan (*Carya illinoinensis*), was the result of nickel deficiency. A commercial product was developed for supplementing nickel and the element has since been recognized as essential for plant growth. Aside from these two taxa, mouse ear disorder has not been documented on other species cultivated in the nursery. As species diversification of managed landscapes becomes a principal issue in the green industry, growers are looking for new crops to produce.

Bitternut hickory (*Carya cordiformis*) is a species gaining the attention of growers, horticulturists, and urban foresters for its horticultural merit. While desirable, one issue encountered by growers is the occurrence of symptoms akin to mouse ear disorder when cultivated in containers with soilless potting substrates. This phenomenon poses a challenge to the development of this species as a nursery crop because hickories are considered difficult to transplant and container production is likely well suited to the adoption of the taxon in landscape horticulture. We questioned whether bitternut hickory is particularly susceptible to mouse ear disorder as a function of nickel deficiency.

Our objectives were to provide evidence that bitternut hickory is susceptible to mouse ear disorder, and to characterize growth responses of symptomatic plants after treatment with the commercial product Nickel Plus® to assess if supplemental nickel ameliorated symptoms.

METHODS

Three-year-old bitternut hickory seedlings grown in a peat-based substrate (#1 containers) were treated two weeks after budbreak with either water (untreated), a substrate drench (37.85ml Nickel Plus®/ 3.79L H_2O), or foliar spray (9.46ml Nickel Plus®/ 3.79L H_2O). A total of 36 plants were used (12 single-plant replicates per treatment).

Plants were grown on a greenhouse bench in Ithaca, NY using a completely randomized design. Data were collected 30 days post-treatment by destructive harvest. Data was analyzed using a one-way ANOVA via JMP Pro version 15.

RESULTS

All untreated controls displayed symptoms of mouse ear disorder, including leaf curling, necrotic margins, rosetting of the stem, suspended leaf expansion, and stem dieback (Figure 1).

Figure 1. Close-up image of symptoms of mouse ear disorder on bitternut hickory.

Symptoms did not manifest on plants treated with either a substrate drench or foliar application of Nickel Plus® and all treated plants resumed normal growth (Figure 2).

Figure 2. Left to right: seedling hickory left untreated (control), treated with a substrate drench, or supplemented via a foliar spray of Nickel Plus®.

ICP-AES test results indicated all untreated controls had undetectable amounts of nickel whereas mean nickel content of plants treated with either a drench or foliar spray comprised ≈2.5 or 83.6 mg•kg^{-1}, respectively. These results indicate symptoms are the result of nickel deficiency, bitternut hickory is susceptible to mouse ear disorder, and that the problem can be corrected by supplementing nickel.

DISCUSSION

Each of the taxa documented as susceptible to mouse ear disorder are ureide transporters. Additional research will screen other woody plants with similar ureide-transporting nitrogen metabolism for susceptibility to mouse ear disorder including other species belonging to the genus *Carya* as well as unrelated taxa.

Acknowledgements: This work is supported by the USDA National Institute of Food and Agriculture, McIntire Stennis/Smith-Lever project 1020775 and the Horticultural Research Institute. We thank Nipan LLC for providing the Nickel Plus® and Johnson's Nursery for sharing their experiences producing bitternut hickory in the nursery

Innovation at Work: Dogwood Breeding at Rutgers University

Erin Pfarr Moreau, Josh Honig, John M. Capik, Thomas J. Molnar

School of Environmental and Biological Sciences, Department of Plant Biology, Rutgers University, New Brunswick, New Jersey 08901 U.S.A.

erin.pfarr@rutgers.edu

Keywords: *Cornus florida, Cornus kousa*, dogwood, genetics, breeding, ddRADseq, genetic diversity, QTL mapping

Summary

At Rutgers University, we are continuing a tradition of innovation as we adopt advanced genetic tools and analyses in our dogwood breeding program. In this paper, we present preliminary results of two studies. The first is a genetic diversity study of 181 *Cornus florida*, *C. kousa*, and interspecific hybrids using the ddRADseq technique. We found that the pink-bracted *C. florida* formed a distinct clade separate from white-bracted trees and were more genetically similar than expected. For *C. kousa*, the accessions separated clearly into two different subspecies groups based on country of origin: ssp. *chinensis* from China and ssp. *kousa* from Korea and Japan. We verified eight previously described ssp. *chinensis* cultivars and found 13 additional cultivars that were previously unknown to be ssp. *chinensis*. We also found 17 cultivars that were genetically intermediate between the two subspecies, indicating they are subspecies hybrids. For both *C. kousa* and *C. florida*, there were also several cases of cultivars that are phenotypically and genetically indistinguishable, representing potential mix-ups in the nursery trade. Our data suggests these cultivars are clones that have been sold under different names in the industry. The largest group of such cultivars contains *C. kousa* 'Satomi', 'Rosabella', 'Schmred' Heart Throb®, 'Hanros' Radiant

Rose®, and 'Grist Mill Pink'. The second study is a Quantitative Trait Loci (QTL) mapping study of *C. florida* to identify regions of the genome associated with powdery mildew (PM) resistance and tolerance that could be used in breeding. Based on 196 full-sibling seedlings of Rutgers H4AR15P25 (PM resistant) x Rutgers H4AR15P28 (PM susceptible), we discovered a QTL on Chromosome 3. This QTL was found to be statistically significant, but explains only 7.8% of the variation in the seedling population.

INTRODUCTION

The Rutgers dogwood breeding program has been innovative since its inception in the 1960s under Dr. Elwin Orton. Dr. Orton pioneered interspecific crosses between the three main species of big-bracted dogwoods, *Cornus florida*, *C. kousa*, and *C. nuttallii*. Over the course of his career, Dr. Orton released 14 dogwood cultivars (Molnar and Capik, 2013).

Figure 1. *Cornus florida* var. *rubra* 'Rutnam' Red Beauty® dogwood.

Today, the program under the direction of Dr. Tom Molnar is focused on breeding powdery mildew (PM) resistant *C. florida* and unique dark pink-bracted *C. kousa* (Molnar, 2018). To support the program, we are using advanced genetic tools to better understand the genetic makeup, pedigrees, and relationships of our breeding selections and cultivars in the industry. We are also using these tools to understand the genetic basis of powdery mildew resistance and tolerance in our *C. florida* breeding program to help more effectively and efficiently breed cultivars with resistance to this disease.

CORNUS FLORIDA AND *CORNUS KOUSA* GENETIC DIVERSITY STUDY

We embarked on a genetic diversity study of 181 *C. florida*, *C. kousa*, and interspecific hybrid accessions. Our study focused mostly on cultivars but also included breeding selections and wild-collected plants. We were interested in answering questions such as: how genetically diverse are dogwood cultivars that are being sold today and in the past? How are these cultivars related to each other and to plants in the wild? Have different subspecies of *C. kousa* been used in breeding and are important ornamental characteristics like pink bract color and variegation specific to certain subspecies or genetic groups? How genetically diverse are the plants in the Rutgers University breeding program? Have our selection efforts narrowed genetic diversity?

Abbreviated methods

We used a technique called double digest restriction-site associated DNA sequencing (ddRADSeq) (Poland et al., 2012) to genotype the plants. The ddRADSeq technique yields thousands more markers than older techniques like SSR markers (simple sequence repeats) and DAF (DNA amplification fingerprinting) that have been used in previous dogwood genetic diversity studies. For the analysis, we used 13,274 markers for *C. florida* and 7,978 markers for *C. kousa*. We are analyzing the data using GBS-SNP-CROP, STRUCTURE, and R programs (Melo et al., 2016; Pritchard et al., 2000; R Core Team, 2020).

Main takeaways so far

For *C. florida*, our data shows that var. *rubra* (pink-bracted) cultivars form a genetic grouping that is distinct from the white bracted plants. This means that they are more genetically similar to each other than expected. This knowledge will be useful as we breed for cultivars with dark-pink bracts and powdery mildew disease resistance while striving to maintain a high level of genetic diversity.

Table 1. *Cornus kousa* cultivar subspecies assignment based on a ddRadSeq genetic diversity study

***Cornus kousa* cultivars**		
ssp. *chinensis*	**ssp. Hybrids**	**ssp. *kousa***
'Autumn Rose'	'Gay Head'	'Akatsuki'
'Big Apple'[a]	'Girard's dwarf'	'Benifuji'
'Blue Shadow'[a]	'KN 144-2' Rosy Teacups®	'Elizabeth Lustgarten'
'Brotzman Dwarf'[a]	'Madame Butterfly'	'Eva'
'China Girl'	'Moonbeam'	'Fascination'
'Flowertime'[a]	'National'	'Gold Star'
'Galzam' Galilean®	'Par Four'	'Grist Mill Pink'
'Greensleeves'	'Primrose Cloak'	'Hanros' Radiant Rose®
'Highland'[a]	'Rochester'	'Kristen Lipka's Variegated Weeper'
'Little Poncho'[a]	'Rutpink' Scarlet Fire®	'Lemon Ripple'
'MADI-11' Mandarin Jewel®	'Snowbird'	'Little Beauty'
'Milky Way'	'Snowboy'	'Rosabella'
'Ohkan'	'Southern Cross'	'Satomi'
'Pam's Mountain Bouquet'[a]	'Square Dance'	'Silver Cup'
'Samzan' Samaritan®	'Summer Fun'	'Schmred' Heart Throb®
'Snow Tower'[a]	'Teddy Scout'	'Summer Games'
'Snowy Peak'[a]	'White Ball'	'Summer Majesty'
'Temple Jewel'[a]		'Summer Stars'
'Trinity Star'[a]		'Tsukuba no mine'
'Triple Crown'		'Weaver's Weeping'
'Tri-Splendor'		
'Wolfeyes'[a]		

[a]Denotes a cultivar that was previously unclassified as ssp. *chinensis*

For *C. kousa*, the accessions were found to clearly group by subspecies origin. Plants collected in China (ssp. *chinensis*) were distinct from plants collected in Japan or South Korea (ssp. *kousa*). Cultivars were identified as ssp. *chinensis*, ssp. *kousa*, or as hybrids of the two subspecies based on their similarity with the wild collected plants of known origins (Table 1). We verified eight previously described ssp. *chinensis* cultivars (Cappiello and Shadow, 2005) and found 13 additional cultivars that were previously unclassified as ssp. *chinensis*. Some authors have written that *C. kousa* ssp. *chinensis* is the superior ornamental form of the species, describing increased vigor, earlier flowering, larger bracts, and excellent fall color (Cappiello and Shadow, 2005; Rehder, 1927). However, it appears this judgement may have been based on a small pool of cultivars and should be revisited with a larger breadth of ssp. *chinensis* accessions.

Most of Rutgers' *C. kousa* breeding selections are ssp. hybrids and form three distinct groups (pink or white bracted) that are relatively closely related. This finding agrees with our breeding program's pedigree information dating back to the 1960s.

Our results also showed that for both C. *kousa* and *C. florida*, variegated plants have arisen spontaneously in different genetic backgrounds and in distantly related plants.

However, all four weeping *C. kousa* cultivars ('Elizabeth Lustgarten', 'Kristin Lipka's Variegated Weeper', 'Lustgarten Weeping', and 'Weaver's Weeping') are closely related to each other, suggesting that the weeping trait has arisen once in the cultivated material and comes from a single source. Additionally, the genetic similarity of 'Kristin Lipka's Variegated Weeper' to 'Lustgarten Weeping' is evidence that 'Lustgarten Weeping' was the source of the original 'Kristin Lipka's Variegated Weeper' sport.

For both species, there are several instances of cultivars that are phenotypically indistinguishable (look the same) that were also found to be genetically identical, suggesting that they are the same plant being propagated under different names (Figure 2).

Figure 2. Four cultivars in our study that are phenotypically and genetically indistinguishable and are likely clones. Photos were taken in the Rutgers dogwood trial in New Brunswick, New Jersey on June 9th, 2019.

Notably, the *C. kousa* 'Satomi' (aka 'Miss Satomi'), 'Rosabella', 'Schmred' Heart Throb®, 'Hanros' Radiant Rose®, and 'Grist Mill Pink' accessions that we included in the study had identical genetic profiles, confirming previous research that these plants represent the same clone (Cappiello and Shadow, 2005; Trigiano et al., 2004). Historical and genetic evidence points to 'Satomi' as the original cultivar of this group.

Currently we are adding more accessions to expand the study and confirm results. These insights will be useful to plant breeders, arboreta, and the industry, as most modern cultivars and popular historic cultivars are represented.

INVESTIGATING PM RESISTANCE IN *CORNUS FLORIDA*

Powdery mildew (PM) caused by *Erysiphe pulchra* is one of the most problematic diseases of *C. florida*. If left untreated in the nursery, PM can decrease growth in stem caliper by 80% and height by 50% in one growing season (Windham et al., 1999). Growers have relied on expensive biweekly fungicide applications since it became a widespread problem in the 1990s (Li et al., 2009). In the landscape PM is rarely treated, but in mature trees it can decrease flowering and overall aesthetic value.

Breeding for resistance has been recognized as the ideal strategy for controlling PM (Li et al., 2009). However, resistance is very rare in natural populations, estimated at 0.1% (Windham and Witte, 1998).

One selection in the Rutgers University dogwood breeding program, H4AR15P25, shows excellent resistance to PM (Molnar, 2018). We are using a technique called QTL mapping to find the regions of DNA where resistance genes may be located. The goal was to learn more about this source of resistance so that we can more effectively use it in our breeding program.

Figure 3. Dogwood powdery mildew

Abbreviated methods

A graphic outline of the methods is presented in Figure 4.

Briefly, to find resistance QTL we crossed the PM resistant tree, H4AR15P25, by a susceptible tree and obtained 196 seedlings for testing. We grew the seedlings in the greenhouse and rated the PM severity in summer 2019, about 1.5 years after sowing. We rated the plants using a 0-100% categorical severity scale (0, 1, 5, 10, 20, 30, ...100). We genotyped the parents and 196 progeny using the ddRADSeq technique and analyzed the raw genetic data with Stacks and JoinMap (Catchen et al., 2011; Van Ooijen, 2006). For the final step of the analysis, we combined the genetic and phenotypic data using the MQM mapping method of MapQTL 6 (Van Ooijen, 2006) to find DNA regions associated with PM resistance.

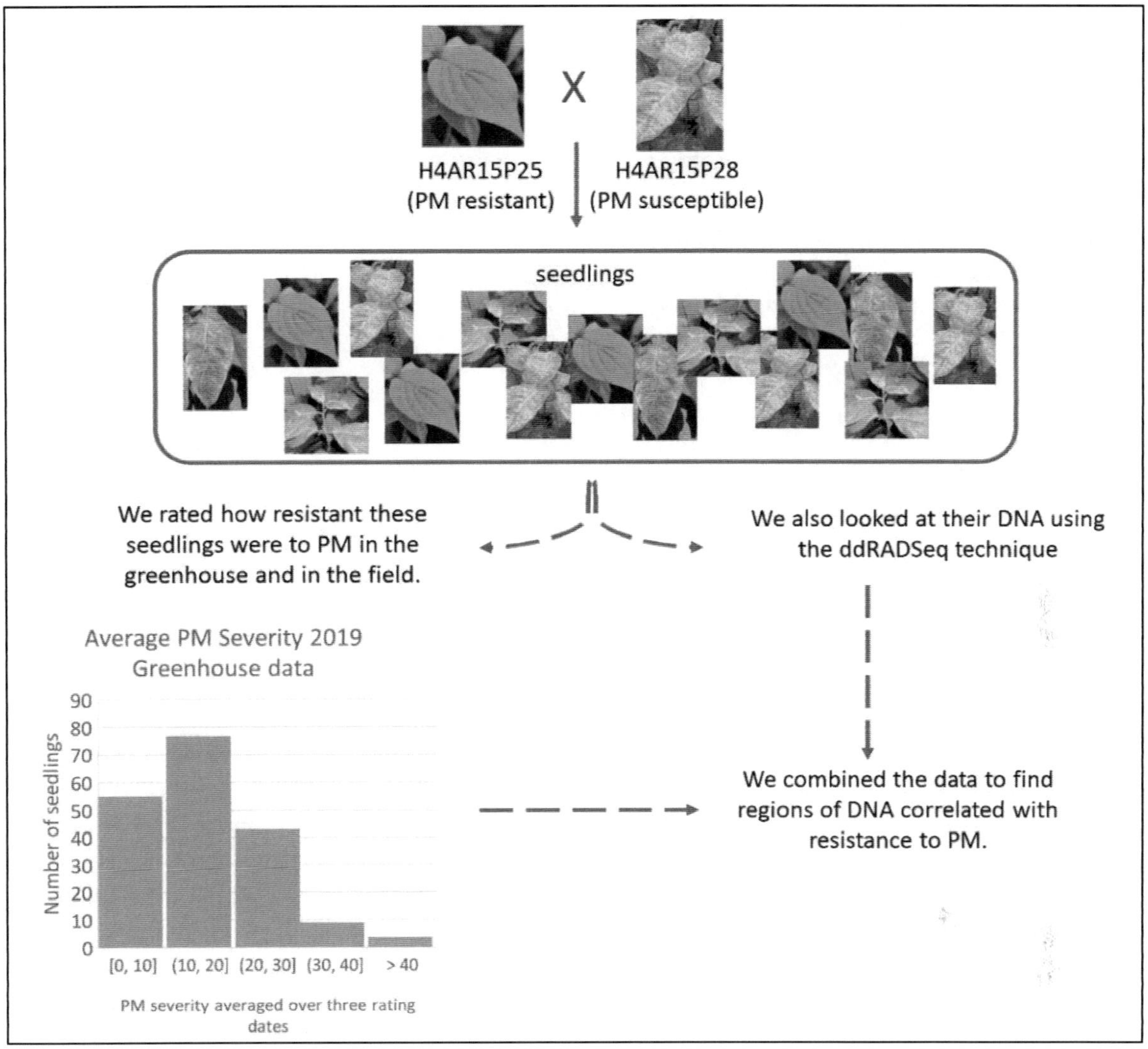

Figure 4. Graphic describing QTL mapping study workflow.

Main takeaways so far

We found one QTL (DNA region) on chromosome 3 associated with PM resistance. It explains 7.8% of the variation we see in the population. This is relatively small, and when taken with the continuous distribution of the PM severity data, suggests that resistance in this seedling population is controlled by many genes with small effects instead of one major gene. Thus, it will be more challenging to utilize this form of resistance in our breeding program; however, we can be more confident that this form of resistance won't break down as fast as when working with a single resistance gene.

The QTL in our study is on a different chromosome than previously discovered QTL (Parikh et al., 2017). These QTL could possibly be stacked to enhance PM disease resistance and breeding efforts are now in progress to combine different, unrelated sources of resistance and tolerance into new breeding populations.

Currently, we are adding to the QTL study by analyzing the results for a related mapping population with 84 individuals—the resistant H4AR15P25 crossed with a different susceptible parent.

Acknowledgements

The authors would like to thank the Molnar lab team for their care of the trees and seedlings in the diversity trial and the mapping populations. We are also grateful for institutions that contributed samples for the genetic diversity study: The Arnold Arboretum, Bernheim Arboretum and Research Forest, The Dawes Arboretum, Morris Arboretum, The Morton Arboretum, The Polly Hill Arboretum, Spring Meadow Nursery, and U.S. National Arboretum. In addition, thank you to Dr. James Leebens-Mack for access to the *Cornus florida* reference genome.

LITERATURE CITED

Cappiello, P. and Shadow D. (2005). Dogwoods: The Genus *Cornus*. Timber Press, Portland, Oregon.

Catchen, J.M., Amores, A., Hohenlohe, P., Cresko, W. and Postlethwait, J.H. (2011). Stacks: building and genotyping loci de novo from short-read sequences. G3: Genes Genomes Genetics *1*:171-182.

Li, Y., Mmbaga M.T., Windham A.S., Windham M.T. and Trigiano R.N. (2009). Powdery mildew of dogwoods: current status and future prospects. Plant Dis. *93*:1084-1092.

Melo, A.T.O., Bartaula, R. and Hale, I. (2016). GBS-SNP-CROP: a reference-optional pipeline for SNP discovery and plant germplasm characterization using variable length, paired-end genotyping-by-sequencing data. BMC Bioinformatics *17*:29.

Molnar, T. (2018). Breeding powdery mildew resistant dogwoods and more at Rutgers University. Intl. Plant Prop. Soc. *68*:385-395.

Molnar, T. and Capik, J.M. (2013). The Rutgers University woody ornamentals breeding program: Past, present, and future. Acta Hortic. *990*:271-280.

Parikh, L., Mmbaga, M., Meru, G., Zhang, G., Mackasmiel, L., Wadl, P., Wang, X. and Trigiano, R. (2017). Quantitative trait loci associated with resistance to powdery mildew in *Cornus florida*. Sci. Hortic. *226*:322-326.

Poland, J.A., Brown, P.J., Sorrells, M.E. and Jannink, J.-L. (2012). Development of high-density genetic maps for barley and wheat using a novel two-enzyme genotyping-by-sequencing approach. PLOS ONE *7*, doi.org/10.1371/journal.pone.0032253.

Pritchard, J.K., Stephens, M. and Donnelly, P. (2000). Inference of population structure using multilocus genotype data. Genetics 155:945-959.

R Core Team (2020). R: A language and environment for statistical computing (Vienna, Austria: R Foundation for Statistical Computing).

Rehder, A. (1927). Manual of cultivated trees and shrubs hardy in North America exclusive of the subtropical and warmer temperate regions. (New York: Macmillan).

Trigiano, R.N., M.H. Ament, M.T. Windham, and J.K. Moulton, J. (2004). Genetic profiling of red-bracted *Cornus kousa* cultivars indicates significant cultivar synonomy. HortScience *39*:489-492.

Van Ooijen, J.W. (2006). JoinMap® 4, Software for the calculation of genetic linkage maps in experimental populations. (Wageningen, Netherlands: Kyazma B.V.).

Windham, M.T., Windham, A.S. and Halcomb, M.A. (1999). Enhancement of growth of flowering dogwood by using fungicides to control powdery mildew. Proc. SNA Research Conference *44*:208-209.

Windham, M.T. and Witte, W.T. (1998). Naturally Occurring Resistance to Powdery Mildew in Seedlings of *Cornus florida*. Journal of Environmental Horticulture *16*:173-177.

Challenges of Improving Urban Forest Canopy in the Chicago Region

Lydia Scott

Chicago Region Trees Initiative, The Morton Arboretum, 4100 IL Route 53, Lisle, Illinois 60532, U.S.A.

lscott@mortonarb.org

Keywords: Tree infrastructure, urban forest, urban forestry, Chicago Region Trees Initiative, tree health, *Quercus*

Summary

Since the completion of the 2010 Tree Census and founding of the Chicago Region Trees Initiative, significant action has been initiated to address many of the challenges identified and to begin to produce desired outcomes. These challenges and outcomes, outlined in the CRTI Master Plan, are being undertaken by a wide range of partners as they become increasingly more aware of the benefits trees provide, the need to expand our regional canopy, reduce threats to our forest, and protect our native oak ecosystems. These challenges are too vast for any one organization to address alone and it will take a wide range of partners and participants, working together, to implement the CRTI Master Plan resulting in a healthy, diverse, and equitably distributed tree canopy benefiting all people in the Chicago region.

INTRODUCTION

Trees and green infrastructure are undervalued in our society partially because their value is not well understood. Urban trees are critical infrastructure. They lower temperatures, improve air quality, reduce flooding, clean water, improve mental and physical health, increase property values, and reduce crime (Turner-Skoff, et al. 2019). The study, the Urban Trees and Forests of the Chicago Region (2010 Tree Census) (Nowak, et al. 2013) revealed that the urban forest in the Chicago region is in a state of "transition". As a result, The Morton Arboretum, along with fourteen other leading organizations from the Chicago region founded the Chicago Region Trees Initiative (CRTI) to begin to address these challenges. The mission of the CRTI coalition, is to *ensure that trees are healthier, more abundant, more diverse, and more equitably distributed to provide needed benefits to all people and communities in the Chicago region.* In 2019, CRTI completed a 3-year effort to gather stakeholder feedback and direction across the seven-county region, resulting in a Master Plan (http://chicagorti.org/MasterPlan) for the Chicago region's urban forest. The plan identified four major goals to address challenges to the urban forest in the Chicago region. They are: Inspire people to value trees, Increase the Chicago region's tree canopy, reduce threats to trees, and enhance oak ecosystems, and CRTI has begun to implement this plan.

DISCUSSION

To drive action and better understand individual and community needs, CRTI has collected one of the largest datasets on urban forestry in the county. This data has been coupled with population vulnerability, air quality, flooding, temperature, and health data. This data is presented as an interactive online resource (http://chicagorti.org/PriorityMap) for all landowners, managers and interested organizations and individuals to inform and prioritize action. In addition, a capacity survey was completed to focus development of resources and trainings based on the capacity of communities to fund and support tree planting and care. The objectives of all of these tools and resources are to improve tree health and capacity to increase quality of life. Additionally, these tools and resources allow for prioritized action, have informed the CRTI Master Plan, and desired outcomes.

Inspire People to Value Trees

The first goal of the CRTI Master Plan is to inspire people to value trees. In order to inspire individuals and communities to value trees we must first understand their values and goals. We must determine how their goals and values can be supported and help them understand how urban trees relate to those goals so they can be inspired to take action. Once people are inspired, ownership and a positive change can take place.

CRTI staff work to engage all levels within the community including elected officials, community administration and staff, volunteers, residents, and community groups. These engagements include discussions of their vision for their communities and development of actions and resources they can utilize, to create change and build excitement in the community for urban trees.

In 2020, The Morton Arboretum, repeated the 2010 Tree Census and found that the canopy had grown from 21% to 23%. The canopy now provides $191 million in annual pollution removal, $3.5 billion in carbon storage, $93 million on carbon sequestration, 1.5 million cubic feet in avoided stormwater runoff, $32 million in energy savings and has a replacement value of $45 billion (Figure 1: Chicago Region Tree Features; Figure 2: Chicago Region Tree Benefits.).

CHICAGO REGION FOREST FEATURES			TOTALS
Number of trees	Chicago	3,997,000	172,297,000
	Seven-County Region	168,300,000	
Tree and shrub canopy cover (i-Tree canopy)	Chicago	16%	23%
	Seven-County Region	26%	
Most common species of trees	Chicago	white mulberry, European buckthorn, tree of heaven	-
	Seven-County Region	European buckthorn, boxelder, black cherry	
Species with the most total leaf area	Chicago	silver maple, Norway maple, white ash	-
	Seven-County Region	European buckthorn, silver maple, black walnut	
Percentage of trees smaller than 6" (15.2 cm) diameter	Chicago	65%	75%
	Seven-County Region	76%	

Figure 1. 2010 Tree Census Canopy (https://mortonarb.org/app/uploads/2021/05/2020-Chicago-Region-Tree-Census-Report__FIN.pdf) – Chicago Region Forest Features.

CHICAGO REGION FOREST FEATURES *continued*			TOTALS
Pollution removal (tons/year)	Chicago	800	18,600
	Seven-County Region	17,800	
Pollution removal ($/year)	Chicago	$36,600,000	$191,600,000
	Seven-County Region	$155,000,000	
Carbon storage (tons)	Chicago	876,500	20,836,500
	Seven-County Region	19,960,000	
Carbon storage ($)	Chicago	$149,000,000	$3,549,000,000
	Seven-County Region	$3,400,000,000	
Carbon sequestration (tons)	Chicago	21,000	542,600
	Seven-County Region	521,600	
Carbon sequestration ($/year)	Chicago	$3,610,0000	$92,610,000
	Seven-County Region	$89,000,000	
Oxygen production (tons/year)	Chicago	23,000	648,800
	Seven- County Region	625,800	
Avoided runoff (cubic feet/year)	Chicago	65,000,000	1,490,000,000
	Seven-County Region	1,425,000,000	
Avoided runoff ($/year)	Chicago	$4,350,000	$99,650,000
	Seven-County Region	$95,300,000	
Building energy savings ($/year)	Chicago	$1,930,000	$32,430,000
	Seven-County Region	$30,500,000	
Carbon avoided (tons/year)	Chicago	1,800	58,800
	Seven-County Region	57,000	
Carbon avoided ($/year)	Chicago	$314,000	$10,094,000
	Seven-County Region	$9,780,000	
Replacement value ($)	Chicago	$2,050,000,000	$44,850,000,000
	Seven-County Region	$42,800,000,000	

Figure 2. 2020 Tree Census, (https://mortonarb.org/app/uploads/2021/05/2020-Chicago-Region-Tree-Census-Report__FIN.pdf) Chicago Region Forest Features (Benefits).

CRTI, also had local LiDAR imagery analyzed enabling interpretation of much of this information at the community and census tract scale. Mapping of the results has been provided to the public on the CRTI website (http://chicagorti.org/PriorityMap) (Figure 3: Chicago Region Prioritization Map). Canopy summary information packets have been developed for each of the 284 individual communities and each of the 50 Chicago wards and 70 Chicago neighborhoods has been provided via the website and shared with community leadership (http://chicagorti.org/interactivemap) (Figure 4: Urban Forestry Summary details). Included within the summary information is a quantification of the value of the communities' trees for air quality, flooding, and carbon storage. These resources support understanding of the value of the urban forest to communities and individuals.

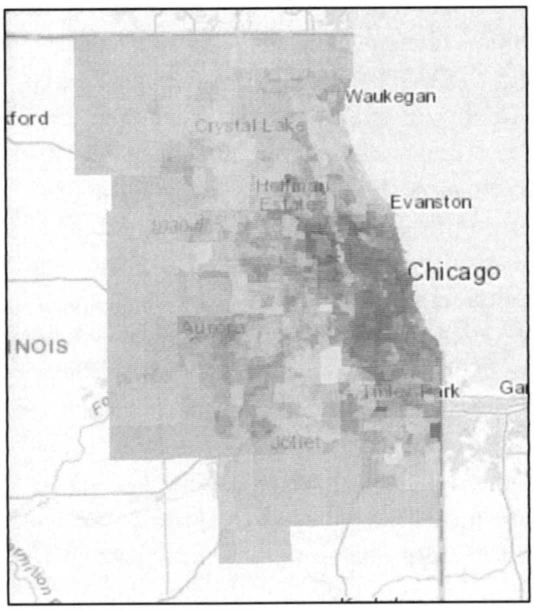

Figure 3. Chicago Region Prioritization Map (http://chicagorti.org/PriorityMap) (the darker the green the higher the priority). The Priority Map combines singular layers (also shown independently) of canopy cover, temperature, air quality, stormwater, and vulnerable populations.

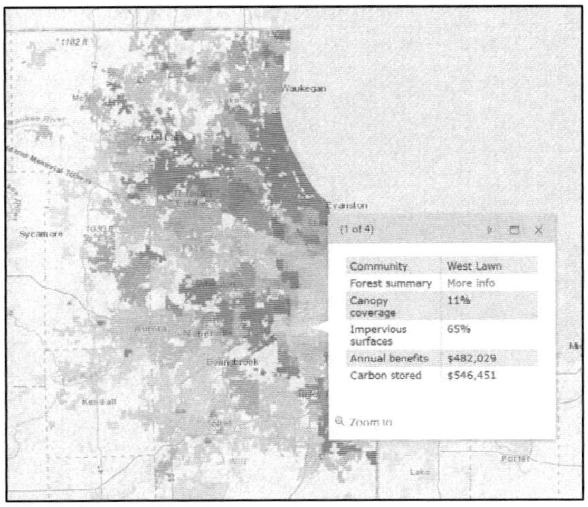

Figure 4. Interactive Community Summary Map, Communities can click on the map to see detailed information about each community or Chicago ward/neighborhood. By clicking on 'More info', Urban Forestry Summary details are presented allowing the community to review a wide range of details about their community's urban forest.

Within each Urban Forest Summary detail packet is a chart that shows tree benefits for three variables — air quality, runoff reduction, and carbon storage. These benefits are calculated using iTree, science-based tools. The chart enables decision makers to weigh their investment in the urban forest with some of the benefits it provides, helping to verify the value of their investment. (Figure 5: Chicago Forest Summary annual benefits for air quality, stormwater, and carbon sequestration.)

In 2020, CRTI launched the Plant Trees (http://chicagorti.org/PlantTrees) digital campaign to help communities and partners share the value of trees with their constituents. The campaign was designed to direct viewers of these outreach pieces to explore deeper messaging and provide increased knowledge and understanding of the value of trees. Key messages include Plant trees to cool and save energy, Plant trees to improve health and well-being,

Plant trees to strengthen communities, Plant trees for nature, Plant trees for clean air, Plant trees for shade and beauty, and Plant trees to help manage stormwater.

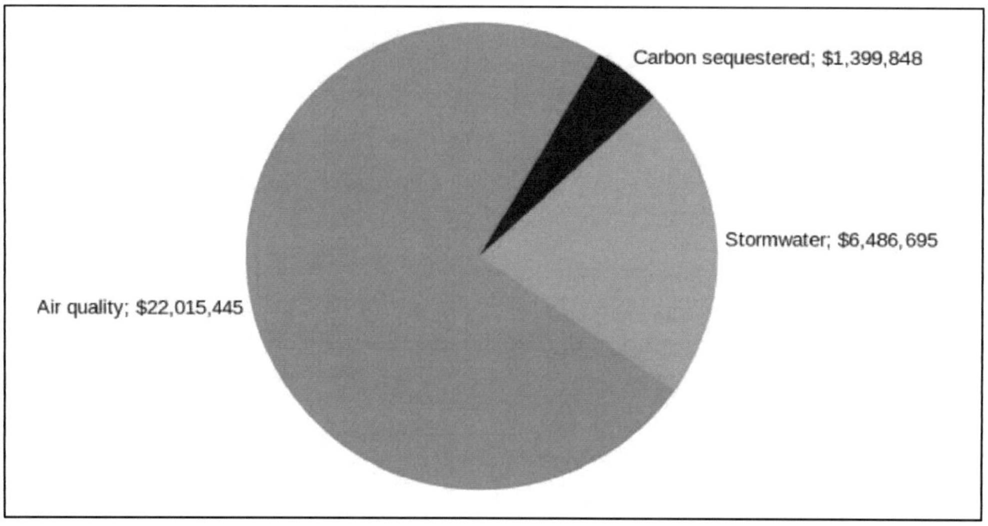

Figure 5. Chicago Urban Forest Summary packet information showing the annual benefits provided by Chicago's trees for air quality, carbon sequestered, and stormwater interception.

Ultimately, CRTI is working to share the value and benefits of trees to the 284 communities, 50 Chicago wards, and 9.4 million people living in the Chicago region so they can make informed decisions about the health of their portion of the urban forest and have a better understanding of its value and services, so they are inspired to preserve, protect, and enhance this critical resource.

Increase the Chicago Region's Tree Canopy

The second goal of the CRTI Master Plan is to increase the Chicago region's tree canopy. In order to achieve and expanded canopy we need to improve preservation and protection of trees through stronger policies, improve their care through increased training and professionalism of tree care, and plant more trees — especially where they are needed most.

Protection and care of existing trees is the greatest need because bigger trees provide bigger benefits. Existing trees need to be protected so they are not removed and can grow to maturity — providing maximum benefits. CRTI, with the help of the Illinois Department of Natural Resources and the USDA Forest Service, is working with the communities to improve their local tree ordinances and policies so that trees are better protected and care for. Funding opportunities are provided for tree ordinances, development of forest master plans, and completion of tree inventories.

The majority of the region's trees are located on private property and CRTI is working to encourage decision makers to implement policies that protect trees on both public and private property – trees are a communitywide asset regardless of ownership. Most communities are reluctant to regulate trees on private property and this is taking some outreach to change. CRTI encourages communities, not ready to implement private property ordinances, to incentivize expanded tree planting and care on private property. This can be accomplished through cost shares for tree purchases, allowing residents to purchase trees through the community's contract at discounted rates and providing assistance in selecting broad species diversity to improve forest resilience. In addition, CRTI assists communities in providing education and outreach to residents through community events, online resources and improve community staff expertise. The CRTI data shows that residential property is the largest land use and has the greatest potential for expanded tree planting and care to increase the region's urban forest. (Figure 6: Urban Forest Canopy summary packet for the Village of Antioch, Illinois. Showing land use and potential plantable space.)

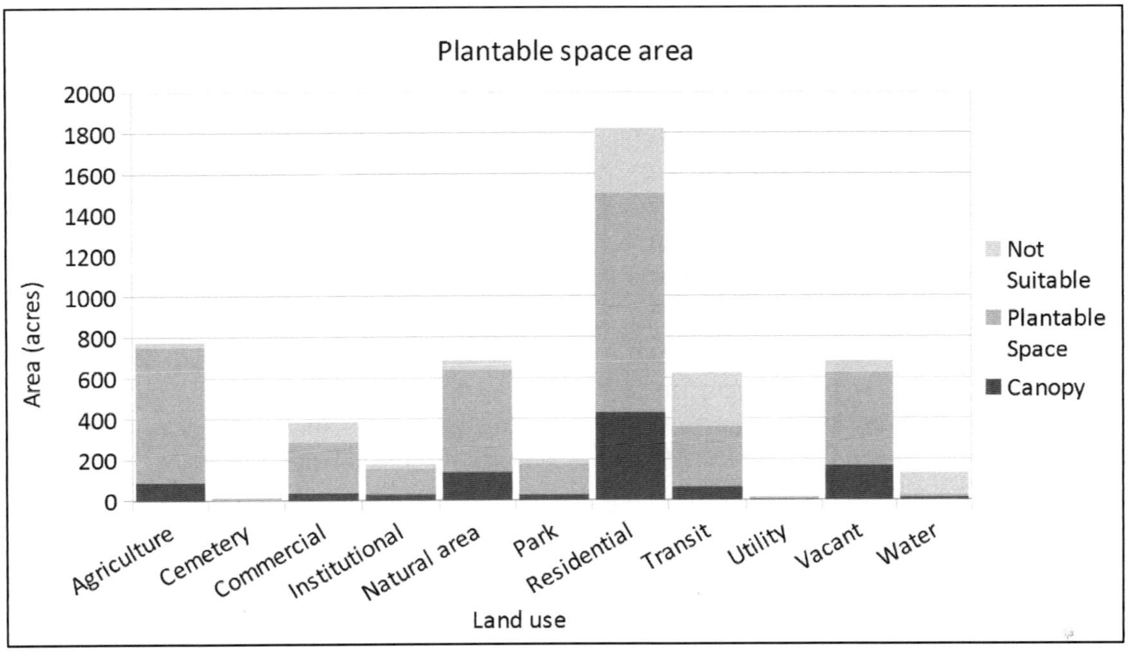

Figure 6. Urban Forest Canopy summary packet table for the Village of Antioch, Illinois. This table shows which land use is the largest in the village and where there is greatest potential for planting to help the village direct its resources to those areas where trees are needed most.

Improved care is important for an expanded tree canopy. Seventy-five percent of all of the region's trees are less than 6 in. in diameter (2020 Tree Census). This is in part because of the enormous percentage of woody invasive species, but also because trees are not living very long. CRTI works with landowners and managers, individually and through workshops and training events, to identify and manage invasive species and their replacement in the landscape, and provides trainings on the importance of professional expertise by using International Society of Arboriculture Certified Arborists. Training is also provided to municipal staff through the Urban Forestry Basic Training and Community Tree Network training sessions. These learning sessions provide opportunities for municipal staff to learn critical basic information on tree planting and care, and chain saw safety. These sessions also provide opportunities for them to network and learn from each other. The end result is to achieve broadened understanding and knowledge about the value of trees, the benefits they provide, and to improve their selection, planting, and care.

Community tree canopy cover ranges from 3% to 66% (2010 LiDAR Analysis) across the region and is often lowest in under-resourced communities.

Historic disinvestment in some communities has led to a below average canopy cover resulting in fewer benefits and services. As discussed earlier, CRTI is prioritizing outreach and resources (using prioritization mapping (http://chicagorti.org/PriorityMap) to focus on those communities that need the most help. CRTI also uses this prioritization to direct federal, state, and private funding sources to purchase trees and provide community-based planting programs, to train communities and citizens to plant and care for trees, and build overall community capacity to support and advocate for trees.

The ultimate goal to increase tree canopy in the Chicago region is based on increasing knowledge, ownership, and advocacy for trees so they are protected, diversity is expanded, and they are planted and cared for correctly so they can grow to maturity.

Reduced Threats to Trees

A range of threats are impacting the current and future health of the urban forest in the Chicago region. Narrow species diversity is a significant concern — especially because of the recent loss of more than seven million ash trees with another six million in decline. Sixty-three percent of the trees in the Chicago region are within 10 species (2020 Tree Census). CRTI recommends not more than 5% of any one species, 10% of any one genus, and 15% of any one family be planted to reduce vulnerability and expand resilience of the forest. Nursery owners have told CRTI that they are limited on the species diversity they can grow because of limited species diversity provided by the liner suppliers. CRTI is working with local nurseries to request expanded species diversity and get trees at smaller sizes (sizes that volunteers and individual property owners can manage easily) through the development of a contract growing program. A contract growing program encourages communities to plan ahead and secure the supply chain by ordering and paying for the species and sizes they need over 5 years.

Another significant threat is invasive species. Forth-five percent of all of the tree species in the Chicago region are invasive species (2020 Tree Census). These species are replacing native species in our natural areas resulting in reduced ecological health and decline of native oak ecosystems. Communities do not manage for invasive species and many private landowners do not know that it is a problem. CRTI has developed resources and hosts workshops on the impact of invasive species. One guide of note is the Healthy Hedges (http://chicagorti.org/healthy-hedges) resource. This is a resource that was developed in poster size for display in nursery centers and also has a brochure that landowners can take with them to their local nursery center. This resource provides guidance on species that would be good replacements for invasive species. (Figure 7: Healthy Hedges poster – a guide to replacements for invasive species.)

Development and human impacts to tree health are another threat. CRTI has developed resources to help communities and landowners understand the impact of construction on trees and to encourage protection and preservation of trees through local policies and incentives. Trees experience threats and can sometimes be threats — in part due to poos maintenance. One such resource is a Tree Risk Toolkit (http://chicagorti.org/ReduceCosts) that includes an important video to educate decision makers.

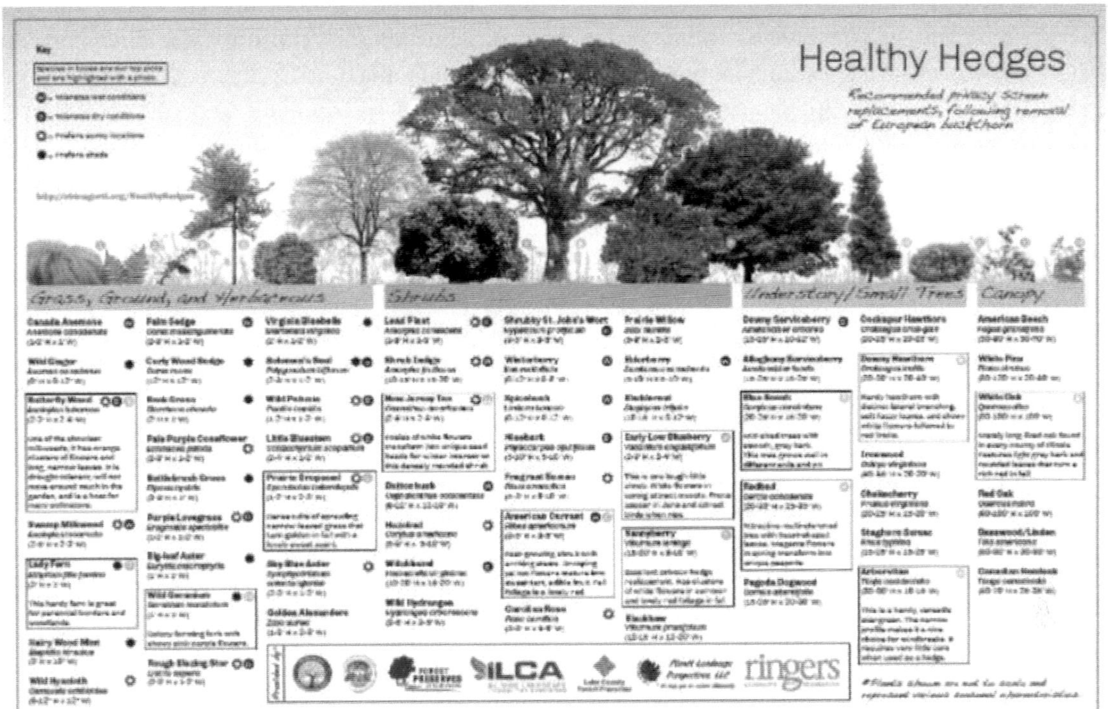

Figure 7. Healthy Hedges poster – a guide to replacements for invasive species.

Climate impacts are creating challenges for urban trees. These challenges include increased inundation, prolonged drought, and severe storm events. The Northern Institute of Applied Climate Science selected the Chicago region as a pilot for identification of community-based climate adaptation strategies and a regional assessment of the vulnerability of the urban forest to climate impacts. Twelve communities and forest preserves participated in this pilot and tools and resources were developed to help other communities and forest preserves in the region.

Enhance Oak Ecosystems

In 2015, Chicago Wilderness published a report on the state of oak ecosystems in the region. This report, the Oak Ecosystem Recovery Plan: Sustaining Oaks in the Chicago Wilderness Region (Fahey, et al. 2015) (http://chicagorti.org/OakRecovery). This plan was based on mapping of remnant oak ecosystems and revealed an 83% loss of these ecosystems. CRTI administers and supports the implementation of this plan through an Oak Ecosystem Recovery Plan Work Group. The group consists of public and private landowners (70% of oak ecosystems are owned by private landowners) with the goal to improve the health of these ecosystems through improved knowledge, practices, and collaboration. CRTI provides funding opportunities and training, through the partnership, to help improve ecosystem health through improved connectivity, protection from extreme browse, reduction of invasive species, and other actions.

Important to supporting preservation and enhancement of oak ecosystems is providing access to local native tree and shrub species. This can sometimes be a challenge as most big-box stores, where many people shop for their trees and shrubs, do not provide locally sourced plant material. Community groups, forest preserves, and conservation organizations work with native nurseries to help distribute their plants to landowners who would typically not have access.

In 2016, CRTI was successful in getting the governor of Illinois to declare October as Oak Awareness Month — OAKtober. Every October, CRTI encourages partners, communities, and individuals to host or participate in events that increase awareness of the need to preserve and protect oak ecosystems, plant more native species, removal invasive species, and work collaboratively with surrounding landowners to reconnect oak ecosystems.

LITERATURE CITED

Fahey, R.T., Darling, L. and Anderson, J. (2015). Sustaining oaks in the Chicago wilderness region oak ecosystems. http://chicagorti.org/OakRecovery

Nowak, D.J., Hoehn III, R.E., Bodine, A.R., Crane, D.E., Dwyer, J.F., Bonnewell, V. and Watson, G. (2013). Urban Trees and Forests of the Chicago Region. https://www.fs.fed.us/nrs/pubs/rb/rb_nrs84.pdf

Turner-Skoff, J.B. and N. Cavender (2019). The benefits of trees for livable and sustainable communities. https://doi.org/10.1002/ppp3.39

Using Apple Bolts to Test Insecticide Efficacy Against Ambrosia Beetles

Zenaida Viloria[1], Raul Villanueva[1], Ric Bessin[2], Win Dunwell[3]

[1]Department of Entomology, University of Kentucky-Research and Education Center, Princeton, Kentucky 42445 U.S.A. [2]Department of Entomology, University of Kentucky, Lexington, Kentucky 40506 U.S.A. [3]Department of Horticulture, University of Kentucky-Research and Education Center-Princeton, Kentucky 42445 U.S.A.

zenaida.viloria@uky.edu

Keywords: Insects, insecticide tests, Fuji apple, *Malus*

Summary

Insecticide tests against ambrosia beetles entails an effective induction of beetle attack in artificially stressed plants, which is laborious and costly. Here we report a 3-year study using 30-cm-long Fuji apple bolts (3-4 cm diam.) to test the efficacy of several insecticides against ambrosia beetles in Lexington and Princeton, KY. Even though, fresh cut bolt technique is not a substitution for trees in ambrosia beetle and insecticide efficacy studies, it facilitates to a large extent the ambrosia beetle research. Pyrethroids and double mode of action insecticides and biopesticides were tested. The pyrethroid ζ-cypermethrin was the most effective for two weeks out of the three pyrethroids tested, followed by λ-cyhalothrin and β-cyfluthrin, whereas the dual mode of insecticide Leverage® (+β-cyfluthrin+ imidacloprid) was tested only in 2020 and was effective for the same period. All insecticide treated bolts were attacked at the of the experiments in 2019 and 2020.

INTRODUCTION

Invasive ambrosia beetles (Coleoptera: Curculionidae: Scolytinae) occasionally cause severe damage to nursery, landscape and fruit trees in spring. In Kentucky, the granulate ambrosia beetle, *Xylosandrus crassiusculus*, is the dominant species that attacks landscape and nursery trees and shrubs (Viloria et al. 2019; Viloria et al. 2021).

Physiological stressed plants mediate the beetle attacks due to the emission of stress volatiles, mainly ethanol. It is very difficult to foresee any ambrosia beetle attack; therefore, preventative application of insecticides is the most appropriate management practice. Artificial induction of ambrosia beetle attacks is needed to test efficacy of insecticide in healthy plants, however the application of flood stress technique (Ranger et al., 2016,) or aqueous ethanol irrigation (Ranger et al.,2018) are arduous and expensive approaches.

The use of ethanol infused bolts (Reding and Ranger, 2020) and bolts with a drilled ethanol reservoir are feasible alternatives to screen insecticides, evaluate damage, identify beetle responsible of attacks, and study insecticide residual effect (Mayfield and Hanula, 2012; Brown et al., 2020; Reding and Ranger, 2020; Jones and Pine, 2018). Thus far, pyrethroids are the recommended insecticides to control ambrosia beetles. The main objective of this study was to assess the efficacy of three pyrethroids, a double mode of action insecticide (pyrethroid + imidacloprid), and two biopesticides against ambrosia beetles using freshly cut apple bolts baited with ethanol.

MATERIALS AND METHODS

Apple bolts

Branches (3-4 cm in diameter) from healthy apple (*Malus domestica* 'Fuji') trees were cut into 30 cm long bolts a day before setting up the experiment. A cavity was drilled at one end of each bolt to make an ethanol reservoir (0.79 cm diam and 4-5 cm deep). The hollowed end was wrapped with sealing film (Parafilm®), thereafter both ends were immersed for few seconds in melted wax to reduce water loss through the cut surface.

Experiments

In 2018 spring, May 9-May 23, a nine-bolt bundle was immersed in Baythroid® XL (ß-cyfluthrin), Mustang® Maxx (ζ-cypermethrin), or vinegars (croton or wood vinegar). In a wooden lot, air dried bolts were hung to trees and suspended about 1m above the ground, keeping a minimum separation distance of about 3 m. After hanging the bolts, three-mL syringe was used to inject 3-4 mL 95% ethanol into the reservoir. Ethanol refill was completed weekly. After ethanol injection the hole was taped to reduce ethanol evaporation. Two control treatments were included: apple bolt with and without ethanol. In 2019, the pesticide test was carried out in Lexington, Fayette Co., Kentucky (April 29-May 20) and Princeton, Caldwell Co. Kentucky (April 23- May 13).

Apple bolts were set in wooden lots, on trees close to the edges. Chemical tested were: ß-cyfluthrin, ζ-cypermethrin, Warrior® II with Zeon Technology (λ-cyhalothrin), hardwood vinegar at 20% and 40%, and control with no pesticide. To facilitate apple bolt deployment and sampling as well as ethanol refill, bolts were hung on a wire that was set at the edge of the woods at approximately 1m above the ground. Every 1m a wire loop was attached to the wire to hang and keep the bolts fixed. Tested insecticides were Leverage®360 (ß-cyfluthrin+ Imidacloprid), ζ-cypermethrin and λ-cyhalothrin and 20% hardwood vinegar.

Chemical concentrations and rates are listed in Table 1. All these insecticides were compared with water plus surfactant as a control treatment.

Table 1. Trade and chemical names and rates of tested insecticides in the spring of 2018, 2019, and 2020.

Treatments	Rates	2018	2019	2020
Baythroid® XL (β-cyfluthrin)	2.8 fl oz/10 gal	X	X	
Mustang® Maxx (ζ-cypermethrin)	4.0 fl oz/10 gal	X	X	X
Leverage® 360 (β-cyfluthrin + imidacloprid)	2.4 fl oz/10 gal			X
Warrior II® Zeon® Tech. (λ-cyhalothrin)	2.56 fl oz/10 gal		X	X
Hardwood vinegar	20, 40%	X	X	X
Croton vinegar	20%	X		
Control	-	X	X	X

Evaluations of Ambrosia Beetle Attacks

Damage caused by ambrosia beetles were recorded as total number entries/bolt, percentage of superficial entries/bolt and percentage of attacked bolts. In 2018, nine bolts per treatment were removed and placed in a bucket (18.93 L), with a net as a lid. Eight weeks later, granulate ambrosia beetles, camphor shot borer and black stem borer were counted (data not presented). In 2019, five bolts/treatment were removed at 10 and 20 days after spray, whereas bolt removal was completed at day 7, 14 and 21 days after spray in 2020. In the last two years, bolts were placed individually in 2 or 4L containers to evaluate beetle emergence (data not presented).

RESULTS AND DISCUSSION

Ambrosia beetle attacks were successfully induced by keeping a permanent source of ethanol emission from apple bolts through a weekly refill of 95% ethanol. In 2018, the experiment was completed in late spring, at that moment the actively flying ambrosia beetle populations were low in western Kentucky (Viloria et al., 2021). However, the numbers of entries/bolt were below 5 in most of the treatments, except Mustang® Maxx, which totally protected apple bolts from ambrosia beetle attacks (Figure 1).

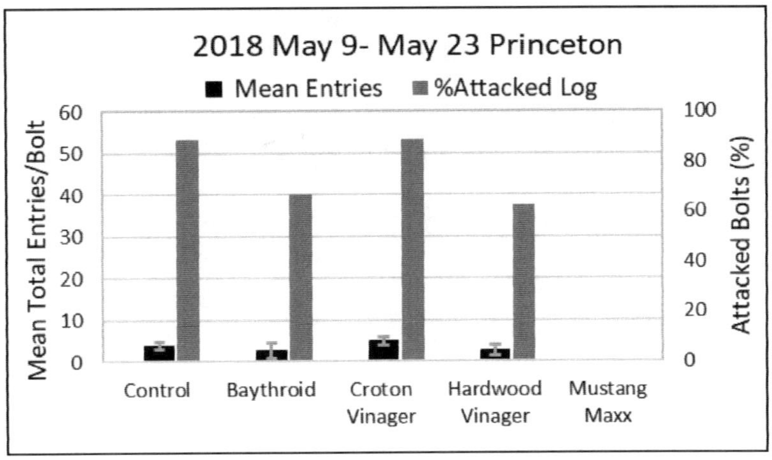

Figure 1. Efficacy of pyrethroid and botanical insecticides in preventing ambrosia beetle attacks to apple bolts in 2018 spring.

Control and croton vinegar showed the highest percentages of attacked bolts (>80%). Vinegar based biopesticides had been reported effective insecticide for a variety of pests (Omulo et al., 2017), nonetheless there is lack of evidence of their effects on borer insects. Neither croton nor hardwood vinegar reduced ambrosia beetle attacks at solution concentrations of 20 and 40%.

The efficacy test completed in two locations in 2019 (Figure 2) showed a considerably higher number of ambrosia beetle attacks for untreated bolts and vinegar treatments at 10 d, and for all treatments at 20 d in Lexington compared with Princeton.

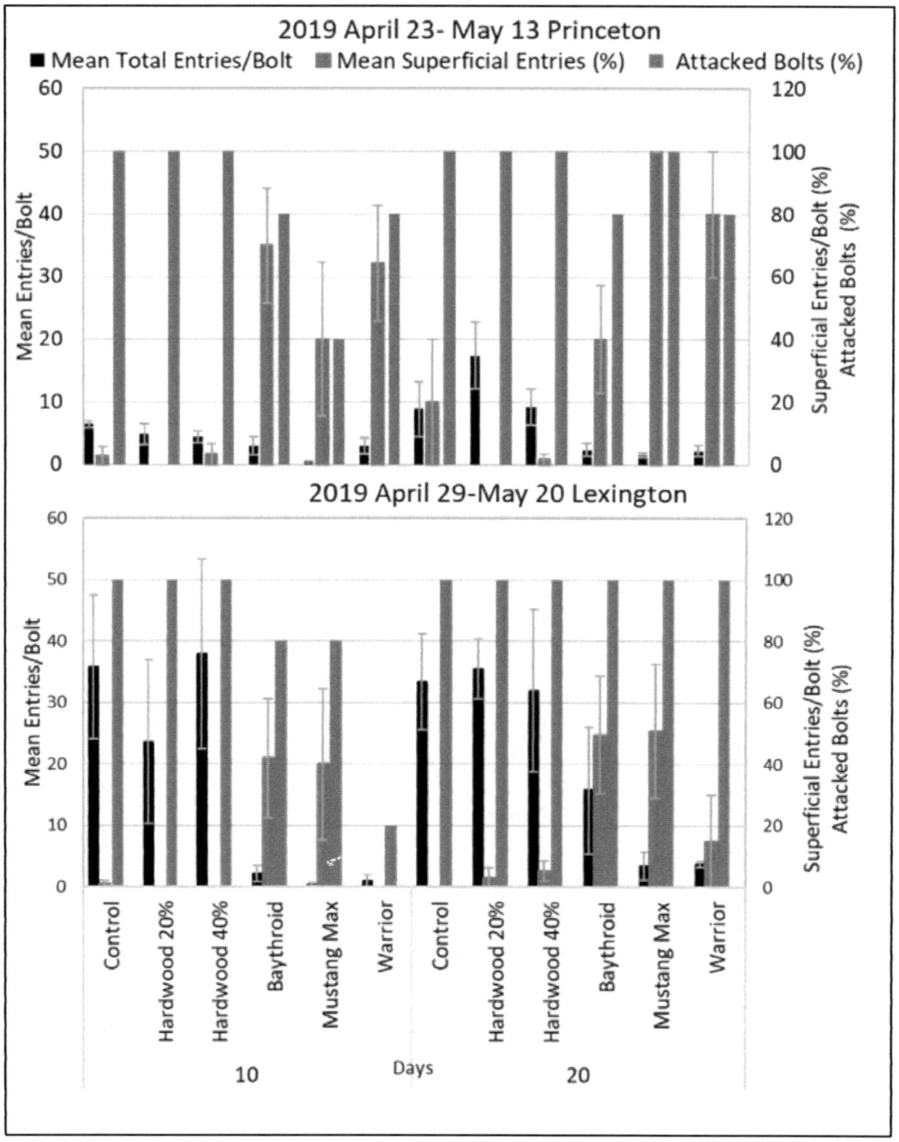

Figure 2. Efficacy of pyrethroid and botanical insecticides in preventing ambrosia beetle attacks to Fuji apple bolts in 2019 spring in Lexington and Princeton, Kentucky.

However, the trend of pesticide efficacy was similar. Pyrethroids reduced significantly the number of entries/bolt. Furthermore, a high percentage of these entries were superficial (<2mm deep). It is likely that tested pyrethroids affect somehow the beetles' capability to bore into the hardwood and stablish a colony in treated bolts. Mustang® Maxx significantly reduced the number of entry holes for 10 d in both locations. This insecticide was still effective in deterring ambrosia beetle attacks for 20 d, at this time its effect was similar to Warrior II® with Zeon Technology. Baythroid XL showed similar effect as non-treated bolts. In a previous study, permethrin, a commonly used pyrethroid against ambrosia beetles, reduced significantly the number of beetle attacks in tree bolts, but did not fully prevent damage (Brown et al., 2020).

The double mode of action insecticide, Leverage® 360, reduced the number of entries/bolt at 7 and 21 d in the 2020 spring (Figure 3); the systemic compound (imidacloprid) of this insecticide might has been curtailed since translocation was interrupted in cut bolts. Imidacloprid reduced of *Euwallacea* sp survival when it was soil drenched (Jones and Paine, 2018). At day 14, Warrior®II showed the lowest % attacked bolts, with similar entry numbers and % superficial entries to those recorded in Mustang®Maxx. At day 21, all apple bolts showed signs of ambrosia attacks, non-treated bolts showed the highest number of entries. Hardwood vinegar had inconsistent results comparing 2019 vs 2020, the number of entries/bolt were similar to untreated bolts in 2019 for the two sites, however in 2020 the number of entries were significant lower in 20% wood vinegar compared to control.

The high incidence of superficial entries suggests a potential use of these insecticides for landscape plants or fruit trees, since the ambrosia fungi are not established, only physical damage remains that may heal to become an undetectable scar. In the nursery crop case, minor damage caused by ambrosia beetles makes trees unmarketable, therefore it is necessary to avoid any attack.

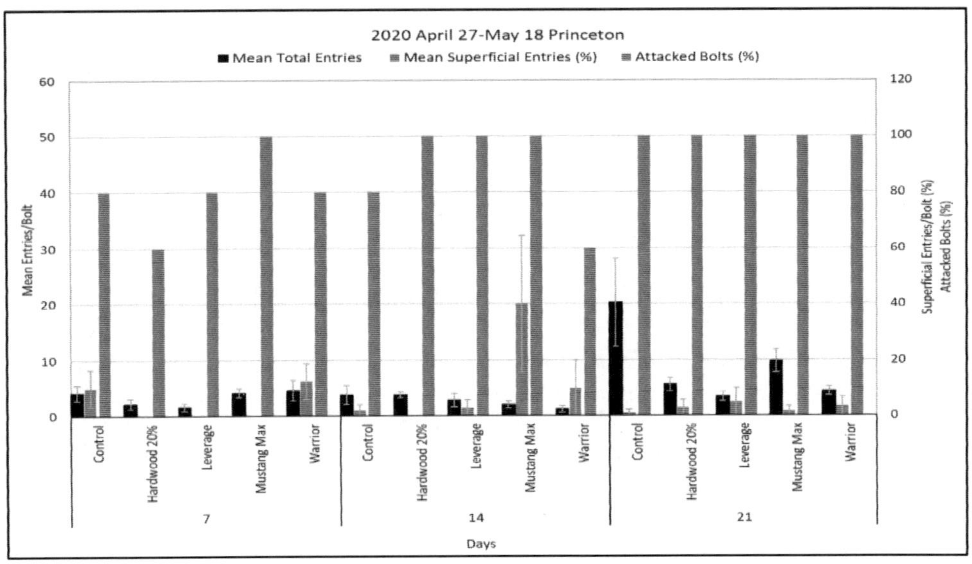

Figure 3. Ambrosia beetle attacks to Fuji apple bolts after a single application of pyrethroids, double mode of action pesticide and hardwood vinegar application in a three-week period of 2020 spring.

CONCLUSIONS

A single application of either ζ-cypermethrin (Mustang®Maxx) or λ-cyhalothrin (Warrior®II) consistently reduced the number of entries for three weeks in low or high ambrosia beetle populations, but the percentages of attacked bolts remained high (80-100%). Whereas β-cyfluthrin (Baythroid®XL) did not show uniform results in Princeton and Lexington in 2019 and Leverage®360 tested only in 2020 was as effective as ζ-cypermethrin and λ-cyhalothrin. Wood and croton vinegars did not deter ambrosia beetle attacks.

Acknowledgements: Thanks to the Kentucky Specialty Crop Block Grant Program-Kentucky Department of Agriculture for providing funds (Award numbers PON2 035 2000000248 in2017 and PON2 035 1800000634 in 2019) for these studies. Also, we acknowledge the involvement of Christine Bradley and the summer interns Rachel Klueppel and Kendell Tobey.

LITERATURE CITED

Brown, M.S., Adesso, K.M., Baysal-Gurel, F., Youssef, N.N. and Oliver, J.B. (2020). Permethrin residual activity against ambrosia beetle (Coleoptera: Curculionidae: Scolytinae) attacks following field aging and simulated rainfall weathering. J. Econ. Entomol. *113*:2418-2426.

Jones, M.E. and Paine, T.D. (2017). Potential pesticides for control of a recently introduced ambrosia beetle (*Euwallacea* sp.) in southern California. J. Pest Sci. *91*:237-246.

Omulo, G., Willett, S., Seay, J., Banadda, N., Kabenge, I., Zziwa, A. and Kiggundu, N. (2017). Characterization of slow pyrolysis wood vinegar and tar from banana wastes biomass as potential organic pesticides. J. Sustain. Dev. *10*:81-92.

Ranger, C.M. and Reding, M.E. (2017). Residue age and attack pressure influence efficacy of insecticide treatments against ambrosia beetles (Coleoptera: Curculionidae). J. Econ. Entomol. *111*:269–276.

Ranger, C.M., Schultz, P.B., Reding, M.E., Frank, S.D. and Palmquist, D.E. (2016). Flood stress as a technique to assess preventive insecticide and fungicide treatments for protecting trees against ambrosia beetles. Insects *7*:40.

Ranger, C.M., Biedermannc, P.H.W., Phuntumarte, V., Beligalae, G. U., Ghoshe, S., Palmquistf, D.E., Muellerc, R., Barnetta, J., Schultzh, P.B., Reding, M.E. and Benz, J.P. (2018). Symbiont selection via alcohol benefits fungus farming by ambrosia beetles. PNAS *115*:4447-4452.

Reding, M.E. and Ranger, C.M. (2020). Attraction of invasive ambrosia beetles (Coleoptera: Curculionidae: Scolytinae) to ethanol-treated tree bolts. J. Econ. Entomol. *113*:321–329.

Viloria, Z. J., Bessin, R., O'Neal, P., Ranger, C., Dunwell, W. and Villanueva, R.T. (2021). Scolytinae in nursery and fruit crops of western Kentucky and seasonal population patterns of four invasive ambrosia beetles. J. Entomol. Sci. *56*:374-386.

Viloria, Z., Bessin, R., Dunwell, W., Travis, G., Ranger, C. and Villanueva, R.T. (2019). Evaluation of pyrethroids and new strategies to deter ambrosia beetle attacks in nurseries. SNA Res. Conf. Proc. *63*:180-186.

Laser-Guided Intelligent Greenhouse Spray System to Deliver Variable-Rate Water, Chemicals, and Nutrients

Heping Zhu[1] and Peter Ling[2]

[1]Agricultural Engineer/Lead Scientist, USDA-ARS ATRU, 1680 Madison Ave, Wooster, Ohio 44691 U.S.A.; [2]Professor, FABE, OARDC/The Ohio State University, Wooster, Ohio 44691 U.S.A

heping.zhu@usda.gov

Keywords: Intelligent sprayers, commercial greenhouse, foliar-applied product, precision technology

Summary

A precision, variable-rate spray system was developed for greenhouse applications. The system utilizes standard greenhouse booms equipped with laser sensors to detect plants and deliver targeted sprays of water, chemicals or fertilizer in a sustainable manner.

INTRODUCTION

The global greenhouse planting area is growing consistently every year. Large commercial greenhouses usually produce various types of plants with various growing sizes in a short period of time to meet the market requirements. Spray applications of pesticides, growth regulators, and nutrients along with irrigation are critical to produce healthy and marketable crops. During the crop production cycles, however, it is very common that small plants are over sprayed, large plants are under sprayed, and empty areas are unnecessarily sprayed, causing significant chemical waste and environment contamination. Thus, precision variable-rate spray technology is needed to improve application efficiency in this controlled environment plant production.

The intelligent spray system offers advantages to deliver variable-rate foliar-applied products precisely. The technology, varying spray outputs with plant architecture and foliage volume, has made significant advances for field orchard crops (Chen et al., 2020). However, due to the environment of a greenhouse being significantly different from that of a field, adapting field equipment to function well in a greenhouse is a new challenge. It would be desirable to extend the already demonstrated variable-rate technology for field applications (Shen et al., 2017) to greenhouse environments economically.

Typical greenhouse operations have the mobile boom travels, at a constant speed, from one end to the other end of a compartment to spray the whole growing area. These over-the-canopy mobile boom sprayers have shown more uniform spray deposits than that of traditional handheld equipment. For the precision variable-rate spraying, detection and characterization of plant is required to spray not only where there are plants but also the right spray volume for plants of different sizes. To reduce the costs of this technology adoption for commercial production greenhouses, minimal modifications of existing greenhouse mobile spray equipment is desirable. Thus, we took advantages of the similarity of using nozzles to manipulate spray outputs and made adaptation of variable-rate spray control systems designed for field sprayers. Our previous analysis has shown a spray control system equipped with a 270° indoor-use laser scanning sensor should be sufficient to control the variable-rate operation with precision (Yan et al., 2018, 2019).

The design of the precision variable-rate spraying technology for greenhouse applications (Figure 1) was based on the field variable-rate system.

Figure 1. The intelligent spray system as a retrofit attached to existing mobile watering booms to deliver water, chemicals, and nutrients with variable rates for greenhouse crops.

The major add-on components of this greenhouse variable-rate boom spray system consisted of a laser scanning sensor to detect objects, a custom designed algorithm to process sensor signals and control spray outputs based on the presence or absence of plants and their structures, an embedded computer, a laboratory-built pulse-width-modulated (PWM) nozzle flow-rate controller, and two 3.6-m long horizontal spray booms. Each boom was equipped with 12 nozzle assemblies coupled with 12 PWM solenoid valves. The two spray booms were attached to a double overhead, mobile rail, watering boom system commonly used in commercial greenhouses. Spray rates discharged were automatically adjusted according to the plant foliage volume. Geometric parameters of the spray system were physical positions of the laser sensor and nozzles, and the sensor travel speed. These parameters were the inputs for the computer program to configure plant structures and locations in complying with acquired laser sensor signals.

The laser sensor was designed for indoor applications with the IP65 protective structure. It generated 1080 detection points at 0.25° angular resolution in a 270° fan shaped plane every 25 ms, and it could detect plants within a 10-m radial range. The distances between the laser sensor and detection points on the plants were acquired through the time-of-flight principle. The plant structure information was collected through continuously scanning the plant from its leading edge to the trailing edge when the spray boom was moving forward. The sensor was mounted in the middle of the two spray booms to access the plant information below the boom.

The embedded computer collected the laser sensor data via an Ethernet interface. The computer managed the spray control system to receive the data acquired from the laser sensor signals, characterize the plant structure, calculate the spray output, and send control commands to activate nozzles. The system performed the variable-rate spray function with the flow-rate controller which consisted of two microcontrollers to control nozzle flow outputs through manipulating duty cycles of the coupled PWM solenoid valves (Liu et al., 2014).

The algorithm for the variable-rate spray system managed the distance data acquired from the laser sensor continuously while the sensor was travelling and converted the distance data into 3-D surface profiles. Each nozzle was assigned to spray a rectangular-shaped section with a given spray width, and was activated to discharge variable-rate outputs based on its corresponding plant presence and sectional canopy volume. In order to ensure entire plants were covered by the sprays, the algorithm was designed to allow the nozzles to start spraying plants at a distance before reaching the plants and stop spraying the plants at a distance after passing the plants. The algorithm was implemented in VC++ programming language.

The system was first validated for its accuracy to synchronize nozzle activation and laser sensor detection of objects and for desired spray volume discharged to the objects. Its performance was then evaluated by quantifying spray coverage inside plant canopies and on the ground at three different travel speeds. The plants, i.e. poinsettias of different species, were at three different growth stages and were placed in different patterns to evaluate the spray system accuracy. The spray deposition, collected using water-sensitive papers, inside the canopies and on the ground were measured and compared. The test results illustrated that spray coverage inside the canopies treated by the spray control system was consistent regardless of the canopy growth stages. Spray coverage inside canopies placed in the continuous placement pattern was greater than that in the group placement pattern, followed by the single plant placement pattern. Effects of travel speed on spray coverage both inside canopies and on the ground were insignificant. Measurements of spray deposition on ground targets at the gaps between plant blocks revealed that the spray control system had the capability to close nozzles in areas with no plants. Additionally, the variable-rate spray system only consumed 21.3% to 89.3% of the spray volume compared to the constant-rate spray mode at different travel speeds.

Our research findings demonstrated the newly developed greenhouse variable-rate spray control system could provide a possibility to increase spray efficiency by greatly reducing spray volume thereby reducing production costs. The system will be further tested for its application accuracy and efficacy under commercial greenhouse conditions and will be used to prevent greenhouse production from excessive waste of water, chemicals and nutrients. Moreover, it will reduce workers from exposure to the harmful chemicals in the confined environment.

LITERATURE CITED

Chen, L., Zhu, H., Horst, L., Wallhead, M., Reding, M., and Fulcher, A. (2021). Management of pest insects and plant diseases in fruit and nursery production with laser-guided variable-rate sprayers. HortScience *56*:94–100.

Liu, H., Zhu, H., Shen, Y., Chen, Y. and Ozkan, H.E. (2014). Development of digital flow control system for multi-channel variable-rate sprayers. Trans. ASABE. *57*: 273-281.

Shen, Y., Zhu, H., Liu, H., Chen, Y., and Ozkan, E. (2017). Development of a laser-guided embedded-computer-controlled air-assisted precision sprayer. Trans. ASABE, *60*:1827-1838.

Yan, T., Zhu, H., Sun, L., Wang, X. and Ling, P. (2018). Detection of 3-D objects with a 2-D laser scanning sensor for greenhouse spray applications. Comput. Electron. Agric. *152*:363-374.

Yan, T., Zhu, H., Sun, L., Wang, X., and Ling, P. (2019). Investigation of an experimental laser sensor-guided spray control system for greenhouse variable-rate applications. Trans. ASABE. *62*:899-911.

New Plant Forum 2021 – Eastern Region IPPS

Kris Bachtell, Moderator

Morton Arboretum, Lisle, Illinois USA

kbachtell@mortonarb.org

Summary

New plants for 2021 are highlighted and described. This year six IPPS-ER breeders presented herbaceous and woody perennial plants.

PRESENTER

Elizabeth Dunham
Knight Hollow Nursery, Inc., 7911 Forsythia St, Middleton, Wisconsin 53562 U.S.A.
liz@knighthollownursery.com

Hydrangea macrophylla **Tuxedo® Series** have deep-purple flushed foliage and occur in both mophead and lacecp form (for different preferences). They are mid-size shrubs (3ft tall × 3ft wide) that flower for 3-4 months from summer to first frost. They can be used in indoor or outdoor containers or also in the landscape. Plants perform best in shade to part-shade but will tolerate some (morning) sun. Available in red and pink. Hardy in USDA Zone 6-9.

Showpiece™ Roses Series are from the breeder of Flower Carpet® Roses. Plants grow up to 2ft wide × 3ft tall. They have full, large, double flowers with a glorious old-world look and a strong pleasant fragrance. Cutting grown (not grafted) plants bloom from late spring to mid fall, are self-cleaning and exhibit excellent disease resistance. They are available in four colors including Berry (dark pink-red), Blush (soft pink), Champagne (creamy orange-peach) and Lipstick (rich reddish-pink). Hardy in USDA Zone 5 – 10.

Lipstick (rich reddish-pink)

Champagne (creamy orange-peach)

Blush (soft pink)

Berry (dark pink-red)

PRESENTER

Brent Horvath
Intrinsic Perennial Gardens, Inc.,
10702 Seaman Rd, Hebron, IL 60034 U.S.A.
brenth@intrinsicperennialgardens.com

***Solidago* 'Sugar Kisses' (Sugar Kisses solidago PPAF)** hybrid *Solidago* has upright woody stems with lime yellow buds in July. Blooms appear August and last into September. Mature plants size is 18 to 24 inches tall by 15 to 18 inches wide with thin foliage that remains clean throughout the season. Plants can be used like a short aster for the front of a mixed border and mix well with short grasses. It is hardy in USDA zones 4 – 9.

PRESENTER

Kim Shearer
The Morton Arboretum, Lisle, Illinois
60532 U.S.A.
kshearer@mortonarb.org

***Acer* 'Morton UW' Morning Starburst™ maple** is a s new maple selection from The Morton Arboretum is a chance seedling selected from our China collection (644-81*1). Originally received as seed from the University of Washington Botanic Gardens as *Acer circinatum* (vine maple) in 1981, today we recognize this tree as a putative hybrid originating from seed of *A. circinatum* and possibly the pollen of *A. pseudo-sieboldianum*.

What makes this selection so special? While there were additional seedlings of *A. circinatum* propagated from the original seed packet shared by UWBG, this is the only individual that has survived the hot and humid summers as well as frigid winters marked by polar vortexes of the Chicago suburbs. It was first noted by Kris Bachtell due to its stunning and consistent fall color and gracefully spreading form. A small specimen tree, this selection has also demonstrated its resilience in commercial nursery production following the 2021 heat wave of the Pacific Northwest region. While others in the field suffered from burned foliage, the Morning Starburst™ selection showed no signs of heat stress.

Fall color

Several clones (grafted onto *A. palmatum* rootstock) were planted in various sites on the grounds of the Arboretum in the fall of 2018 just before the record-breaking 2019 polar vortex. While many of the trees on Arboretum grounds suffered from dieback and decline due to the unprecedented low temperatures, the freshly planted Morning Starburst™ have shown no signs of stress due to cold damage. A low branching small tree with a height of approximately 20-30 ft at maturity and a spread of approximately 15-20 ft, USDA hardiness zone 5 – 8.

Available from: Kuenzi Turf & Nursery, Heritage Seedlings & Liners, and J. Frank Schmidt & Son Co.

***Gymnocladus dioicus* 'Morton', Skinny Latte® Kentucky coffee tree** is an upright and columnar male selection of Kentucky coffee tree is a chance seedling discovered in The Morton Arboretum collections. Originally accessioned as a seedling purchased from Cole Nursery in 1958, the first clonal propagation of this selection took place in 1968. Four of the original clones currently growing in the Arboretum tree breeding nursery are pictured in the photo below.

This selection has tight upright branching forming a narrowly fastigiate habit, and there has been no evidence of branch sports producing female flowers and fruit. Will make a great street tree and can provide some architectural interest in a winter landscape. Height 50', Width 15-20', USDA Hardiness zone 3-8

Summer

Winter

PRESENTER

Shelby French and Sam Hoadley
Mt. Cuba Center, 3120 Barley Mill Road,
Hockessin, DE 19707 U.S.A.
sfrench@mtcubacenter.org
shoadley@mtcubacenter.org

Mt. Cuba Center conducts trials on native species and their related cultivars to determine their horticultural merit and ecological value in the mid-Atlantic region. The Echinacea trial, which included 75 cultivars and species, was evaluated over three growing seasons from 2018-2020.

Why trial *Echinacea*?

- Evaluate new cultivars in the mid-Atlantic for horticultural appeal
- Re-evaluate a selection of previous top performers from Mt. Cuba Center's first trial of the genus Echinacea conducted in 2007-2009.
- Observe plants for disease resistance and longevity
- Record pollinator preference (bees, wasps, and butterflies)

***Echinacea purpurea* 'Pica Bella' (5.0)** was a top performer in our latest trial and tied for first in our original trial of the genus. It is vigorous and uniform and a more compact selection compared to the species. Ranked in the top 15 cultivars for pollinator attraction.

***Echinacea* 'Sensation Pink' (4.9)** was bred by Marco van Noort in the Netherlands. It has incredible neon pink blooms on contrasting dark stems. It ranked in the top five cultivars for pollinator attraction.

***Echinacea* 'Santa Fe' (4.8)** is a Proven Winners® introduction and is also known as *Echinacea* Lakota® Fire in the trade. It was the top rated red/orange cultivar in the trial. Minor variations between plants but, overall, they formed tidy 2' mounded plants with exceptional floral displays. A top-five pollinator visitor favorite in 2018 and 2019.

***Echinacea* 'TNECHKR', KISMET® Raspberry (4.7)** is one of several Terra Nova Nurseries introductions that are included in our top performers list. It is a vigorous semi-compact plant that has oversized saturated pink blooms and extra wide petals creating tremendous floral impact.

PRESENTER

Angela Treadwell-Palmer
Plants Nouveau, Box 40125,
Mobile, Alabama 36640 U.S.A.
angela@plantsnouveau.com

***Echinacea* Fine Feathered™ 'Parrot' PPAF** originates from AB-Cultivars, Netherlands. It has bright, golden yellow petals with a pumpkin orange halo surround a velvety brown cone for a stunning combination. Not only is the color brilliant, but the number of flowers is also astounding. Each retail-ready plant can have up to 50 flowers per pot. That's amazing!

Plants are 18-24 in. tall by 24 in. wide. Use in borders, containers, cutting gardens, foundations, mass plantings and urban gardens. Hardy in USDA zones 6-9. They are tissue culture propagated and available for purchase from TC - AB-Cultivars; Liners - Garden World, Creek Hill, Jardins de Paquette.

***Leucothoe axillaris* 'ReJoyce' PP#30821** from Greg Joyce of Edgar Joyce Nursery Plants grow 24-36 in. tall by 36-42 in. and offer year-round interest. Beginning in spring, new leaves erupt in a blaze of red over tidy green mounds. White, urn-shaped flowers form in short clusters on gently arching, yet dense branches, adding a graceful elegance to any setting, whether it be container plantings or your favorite garden path. As summer begins to fade, 'ReJoyce' ignites again with the entire plant bursting into autumn colors of deep wine red- persisting through winter. Re-Joyce is also a grower's dream because cuttings can be rooted throughout the growing season, the plants do not get leaf spot in the nursery and they are much more resistant to root rot. Use in borders, foundations, woodland gardens, and slopes. Hardy in USDA zones 6-9. Plants from vegetative cuttings are available for purchase from Manor View Farms, Edgar Joyce Nursery, Overdevest Nurseries, Saunders Brothers, Piedmont Carolina Nursery.

Summer color

Winter color

***Metasequoia glyptostroboides* 'Soul Fire' PP#32580** originates from Andrew Schenck. This new gold needled redwood was found by one of our favorite plant geeks, Andy Schenck, owner of Sam Brown's Nursery. It has extraordinary spring, summer and fall color that will light up any garden space. Each spring, bright, lime-green needles emerge with a rosy-orange frosting, making for a two-toned effect. As the summer days grow longer, and the heat and humidity pick up, the needles change to bright, chartreuse and don't fade until fall brings on a bright, orange hue. Plants can grow 15-18' tall by 10-15' wide. Soul Fire can take full sun, but will be just as colorful in part shade. Dawn redwoods are also good for urban sites where temperature and soil moisture extremes can often be extreme. Use as a specimen focal point, urban gardens, or wet areas. Hardy in USDA zones 4-9. Grafted liners are available for purchase from Hans Nelson and Sons.

PRESENTER

Todd West
North Dakota State University,
Plant Sciences, NDSU Dept 7670, PO Box 6050,
Fargo, North Dakota 58108-6050 U.S.A.
todd.p.west@ndsu.edu

***Rhododendron × kosteranum* 'FireDak', Fireflare Orange® mollis azalea** is a single plant selection originating from a hybrid Mollis azalea seedling population. It has performed admirably for over 40 years of evaluation. This NDSU deciduous hybrid azalea cultivar has proven to be pH adaptable and winter hardy. It is a dense, compact semi-dwarf shrub with a medium growth rate to four feet tall and five feet wide. The deciduous foliage is a medium green summer color with outstanding yellow orange to reddish-purple autumn color. USDA hardiness zones 3b – 8. Propagation is from stem cuttings or tissue culture (microcuttings will be available Spring 2022 from Mountain Shadow Nursery, Washington).

Flowering plant

Fall color

***Betula tianshanica* 'EmerDak', Emerald Flare® Tianshan birch** is a distinctive narrowly pyramidal form. Attractive white exfoliating bark with gray and slight orange undertones. Cold hardy and tolerant of higher pH soils (>8.0 pH) and drought. It has an upright, pyramidal habit with a medium to fast growth rate to a height of 30 feet and 12 feet wide.

Summer foliage is of high quality without blemishes resulting from birch leafminer or leaf spot. During summer drought conditions, no foliar stress symptoms such as leaf scorch or early leaf drop which is seen on many other birch species. Fall color is a bright golden-yellow. Hardy in USDA zones 3b – 6. Tissue culture plants are available from Baker's Nursery (Ontario, Canada), Knight Hollow Nursery (WI), Evergreen Nursery (WI). Birch tissue culture liners typically do not grow 1st year after transplant and will put on growth 2nd year; Emerald Flare® starts growing 1st year.

Pinus mugo **'HyDak', Hyland Splendor® mugo pine** is a single plant selection originating from a population of *Pinus mugo* seedlings. The selection has darker green foliage and maintains this high quality dark green needle color during the winter months as compared to the other sibling trees from the original seedling population. It has an upright, pyramidal habit with a slow to medium growth rate to 15 to 18 feet tall and 8 to 10 feet wide.

Nursery trials have shown that Hyland Splendor® grows 20% faster in a controlled commercial nursery production setting as compared to 'Tannenbaum'. This increased growth rate will increase profitability with growing this selection. The species and prominent cultivar, 'Tannenbaum', often develop an undesirable yellowish-green winter needle color. Hyland Splendor® is unique in that it doesn't develop this typical winter needle color and stays a superior dark green. USDA hardiness zone 3a to 7.

Form

Standard *Pinus mugo* on left and Hyland Splendor® on right

PROCEEDING'S PAPERS

SOUTHERN REGION OF NORTH AMERICA

Dr. Fred T. Davies, Jr., Regional Editor

Forty-fifth Annual Meeting - 2021

Mobile, Alabama USA

Technical Sessions of International Plant Propagators' Society-Southern Region of North America (SRNA) Annual Meeting

Brie Arthur

Brie Grows, B.A. Communications

7624 Troy Stone Drive, Fuquay-Varina, NC 27526, USA

brienne.gluvna@gmail.com

Keywords: Annual Meeting, Southern Region of North America (SRNA)

Summary

The 45th Annual Meeting of the International Plant Propagators' Society-Southern Region of North America (IPPS-SRNA) convened at 8:00 pm on 25 October 2021 at the Renaissance Mobile Riverview Plaza Hotel, Mobile, Alabama, with President Brie Arthur presiding.

President Brie Arthur

President Arthur welcomed everyone to Mobile, Alabama for the 45th Annual Meeting of the SRNA. It is so awesome to be here in Mobile to "seek and share" with one another! The relationships and experiences forged at our meetings are what distinguishes the IPPS. She remarked on the challenges of the past two years: cancelling the 2020 meeting in Tulsa, Oklahoma because of the Covid-19 pandemic. In place of the annual meeting, the SRNA successfully led, planned and executed the 3-day North American Virtual Summit (NAVS) during Oct 27-29, 2020. There was a total of a total of 947 participants on the zoom from 15 countries.

She thanked Local Site Committee Chair, Dr. Jeremy Pickens and his committee and volunteers for their outstanding work in arranging the excellent tours, hotel, other planning activities and all their attention to detail. Because of the hotel's Covid restrictions, attendance was limited to 150 registered participants by the conference hotel, however there are 161 registered for the conference – and many who wanted to register were turned away. The SRNA was one of the few international regions to meet this year and conduct tours.

She acknowledged Sec-Tres, Donna Foster, for her outstanding service. The financial status of the SRNA remains strong.

Arthur thanked the Executive Committee, and the Sponsorship Committee of Tom Saunders, Bobby Green and Judson LeCompte who raised $54,500 – which was outstanding with the challenges of the pandemic! Arthur encouraged the membership to thank, visit and show their support of our sponsors during the meeting. She encouraged all members to make new members and first-time attendees feel welcome — share with them and seek from them. She called for good questions and enthusiastic participation at the Tuesday night question box & ice cream social. Two questions posed to attendees were: 1) What idea/approach has improved your bottom-line most over the course of your profession? and, 2) What is the greatest challenge that you are facing as a green-industry professional?

Arthur announced that the SRNA is in its fourth year of the Southern Region Educational Endowment, with a base donation of $20,000 from an anonymous donor. The Education Endowment balance is now $96,250 – and growing. It will greatly enhance our region's ability to support students and early career professionals – and ensure continued quality of the outstanding educational programs our region is known for. At tonight's banquet, there will be the awards program and recognition of new members - followed by the live auction. All of this year's contributions to the silent and live auction are to go to the Endowment Fund – so please contribute! She thanked Kevin Gantt for leading the endowment effort. This year the SR-IPPS initiated the *Margie Jenkens Industry Scholarship* to support industry professionals attending our conference for the 1st time.

Because of Covid, the *Early-Career Propagator Exchange* program between the SRNA and the European Region was put on hold this year, but will resume in 2022.

This is the tenth year the SRNA is doing the *Vivian Munday Young Horticultural Professional Scholarship Work Program (Vivian Munday Scholarship).* We currently have two young professionals: Kayla Morrison of Oklahoma State University and Teagan Young of the University of Florida - who are assisting Sec-Tres Donna Foster – and making a strong contribution to this year's program.

Arthur thanked Program Chair and 1st Vice-President, Bobby Green for the excellent program and world-class speakers he assembled – two years in a row! Because of the pandemic challenge, the 2020 meeting in Tulsa, Oklahoma was cancelled, and Green had to assemble a new slate of speakers for the 2021 Mobile meeting

Program chair Bobby Green

Program Chair Bobby Green welcomed all members, guests and students. He acknowledged President Arthur for her leadership and very capably serving as President during the demanding two past years! He thanked the membership for the opportunity to serve them, and then reviewed the scheduled program. The Question Box, scheduled for Tuesday evening, was to be moderated by Dr. Judson LeCompte. He then introduced the first moderator, Christine Coker from Mississippi State University.

Figure 1. President Brie Arthur (left) with Bobby Green (right), Program Chair of the 2021 Mobile, Alabama, 45th annual conference.

"To Eat, or Not to Eat, That is the Question" - Answered by Real-Time Monitoring Techniques Combining with Computational Analysis for Feeding Behavior Study of Crapemyrtle Bark Scale (*Acanthococcus lagerstroemiae*)

Bin Wu[1a], Elizabeth Chun[2], Runshi Xie[1], Gary W. Knox[3], Mengmeng Gu[4], and Hongmin Qin[2]

[1]Department of Horticultural Sciences, Texas A&M University, College Station, TX 77843, USA; [2]Department of Biology, Texas A&M University, College Station, TX 77843, USA; [3]Department of Environmental Horticulture, University of Florida/IFAS North Florida Research and Education Center, Quincy, FL 32351, USA; [4]Department of Horticultural Sciences, Texas A&M AgriLife Extension Service, College Station, TX 77843, USA.

hqin@bio.tamu.edu; mgu@tamu.edu

[a]First Place – Charlie Parkerson Graduate Student Research Paper Competition

Keywords: Data mining, electrical penetration graph, sap-sucking hemipteran, stylet penetration

Summary

Crapemyrtle bark scale [(CMBS) *Acanthococcus lagerstroemiae*], an invasive and polyphagous sap feeder, has spread across 17 U.S. states. The infestation of CMBS negatively impacts the flowering and fruiting of various ornamental and fruit plants. Crapemyrtle bark scale host confirmation is critical to determine the insect's potential risks to the Green Industry and help develop strategic management of CMBS. Previously confirming CMBS hosts was time-consuming. We investigated the CMBS feeding activities using the electrical penetration graph (EPG) to monitor real-time stylet penetration to determine potential hosts more efficiently. First, we characterized typical EPG waveforms (waveform C, waveform potential drop, waveform E and waveform G) of feeding activities for CMBS on a validated host, *Lagerstroemia limii*. We then tested the feeding behavior of CMBS using different species, including *L. speciosa*, *L. indica* × *speciosa* '18096', Mexican beautyberry (*Callicarpa acuminata*), three Ficus species (*F. pumila*, *F. tikoua*, and *F. auriculata*), and soybean (*Glycine max*), with the positive control (*L. limii*). Results showed that plant species significantly impacted phloem sap ingestion of CMBS, which could be used to rapidly confirm a potential CMBS host.

INTRODUCTION

Crapemyrtle bark scale (CMBS), *Acanthococcus lagerstroemiae* (Hemiptera: Eriococcidae), is an invasive polyphagous insect (Kozár et al., 2013) which has spread across 17 U.S. states since its initial report in Texas in 2004 (EDDMapS, 2021). The reduction in flowering or fruiting on ornamental plants and crops resulted from the infestation and the observation of CMBS found on native species sharpened the concern about this invasive insect's threat potential to the Green Industry and ecosystems (Gu et al., 2014; Merchant et al., 2018; Wu et al., 2021; Xie et al., 2020; Zhang and Shi, 1986). Therefore, it would be crucial to determine the host range of this relatively new invasive insect for better estimating its risks to the local economy and ecosystem.

The host range assessment involves accepting or rejecting plant species via insect feeding performance (Schoonhoven et al., 2005). However, measuring the feeding performance of sap-sucking insects typically needs time-consuming tests regarding biological traits (Herbert et al., 2009; Wang et al., 2019; Wu et al., 2021; Wu et al., 2010). Stylet penetration could be a vital parameter for sap feeders to rapidly assess the host range through a real-time feeding monitor technique and an electrical penetration graph [EPG (Prado and Tjallingii, 1997)]. The EPG technique can track the position of the hemipterans' stylet tips in different plant tissue via voltage fluctuations amplified as specific EPG waveforms (Tjallingii, 1985), and these EPG waveforms were associated with biological feeding activities through histology correlation work (Tjallingii and Esch, 1993). Applying the EPG monitoring techniques in the feeding behavior study of CMBS could confirm the host rapidly and improve the understanding of the CMBS-plant interaction, which would further assist in developing integrated management of CMBS.

To date, little is known about the feeding behavior of CMBS or the CMBS-plant interaction. This study aimed to characterize EPG waveforms related to the feeding activities of CMBS on a validated host plant (*Lagerstroemia limii*) and assess the plant host suitability for CMBS rapidly by comparing its feeding parameters among different plant species.

MATERIALS AND METHODS

Insects and Plants. Colonies of CMBS were established by attaching CMBS-infested branches to healthy *L. limii* plants and maintained in a handmade chiffon mesh-covered cage (58.0 cm long × 58.0 cm wide × 50.0 cm high) in a CONVIRON® (Controlled Environments Ltd., Winnipeg, Manitoba, Canada) growth chamber [25 ± 1 °C, 60±5 % relative humidity (RH), and a photoperiod of 16 h light

(L):8 h dark (D)] at the Department of Biology, Texas A&M University. All CMBS used for EPG recordings were female adults of CMBS (2.1 ± 0.7 mm long; 1.2 ± 0.5 mm wide) obtained from the colony.

Plants used for characterizing the feeding behavior of CMBS were the confirmed host plant *L. limii* (n = 20). Plants used for comparing the feeding parameters by plant species were *L. limii* (n = 25), *Lagerstroemia speciosa* (n = 25), *Lagerstroemia indica × speciosa* '18096' (n = 20), *Callicarpa acuminata* (n = 20), *F. auriculata* (n = 20), *Ficus pumila* (n = 20), *Ficus tikoua* (n = 20), and *Glycine max* (n = 25). The Arabic number in the paratheses represented how many plants were tested for each species. The crapemyrtle plants (*L. limii* and *L. speciosa*) were initially provided by North Florida Research and Education Center (Quincy, FL). The crapemyrtle hybrid '18096' was selected from our crapemyrtle breeding program at the Department of Horticultural Sciences (College Station, TX). The *Ficus* species and Mexican beautyberry (*C. acuminata*) were initially provided by John Fairey Garden Conservation Foundation (Hempstead, TX). All these test plants were propagated via cuttings. They were maintained in 1 qt plastic pots (The HC Companies, Twinsburg, OH) filled with Jolly Gardener Pro-Line C/25 growing mixture (Oldcastle Lawn and Garden Inc, Poland Spring, ME) in the greenhouse at 25 ± 5 °C, 50 ± 10% RH, and a photoperiod of 10.5:13.5 (L:D) h.

Electrical penetration graph recordings of CMBS feeding on different plant species

The CMBS penetration activities were monitored by the EPG devices on different plant species, using individual CMBS female adults in a Faraday cage to characterize the feeding behavior of CMBS and test if plant species affect the feeding behavior. The EPG experiment was conducted in a climate-controlled room (25 ± 1 °C, 60±5 % RH, and a 16 h: 8h photoperiod) at the Department of Biology. The feeding behavior was monitored and recorded for 24 hours, and the recording was replicated using a new insect and a new plant for each species per time.

All typical EPG waveforms in the recordings were labeled manually. After comparing EPG waveforms of other sap-sucking insects (Prado and Tjallingii, 1994; Tjallingii, 1985; Tjallingii and Esch, 1993; Tjallingii, 2006), typical feeding waveforms of CMBS on the host were characterized with the help of EPGminer, a semiautomatic analysis application. Based on the biological feeding activities, EPG parameters about the feeding activities of CMBS on each plant species were compared, including the total duration of stylet pathway phase (waveform C and potential drop), total duration of phloem salivation (E1), total duration of phloem ingestion (E2), and total duration of xylem ingestion (G).

Data processing and statistical analysis

The ggplot2 (Hadley, 2016) and the plotly (Sievert, 2020) were used to generate visuals from R. The EPGminer package (supplementary) was newly developed to extract and analyze the EPG data. The values (mean with standard deviation) for the frequency and voltage (relative amplitude) were calculated by using functions, wave_topfreq and wave_volts, respectively.

Data analysis was performed using JMP® 16 (SAS Institute, Cary, NC). The parameters listed in Table 1 were analyzed using the one-way analysis of variance (ANOVA) to test the effect of plant species on the total duration of each feeding waveform. Tukey's Honestly Significant Difference (HSD) test (α =0.05) was used to compare the difference in each mean value.

Table 1. Characteristics of the EPG waveforms recorded during CMBS feeding on *Lagerstroemia limii*

EPG waveform		Waveform characteristics				Correlations
		Voltage level	Frequency (Hz)		Relative amplitude (%) [Z]	Activities assigned for similar waveforms in other hemipterans
			Min-Max	Medium ± SE		
C		Extracellular	0.59-1.61	0.98 ± 0.10	11.81± 1.00	Sheath salivation and other intercellular stylet pathways
pd	pd1	Intracellular	0.42-6.10	4.35 ± 0.71	20.20 ± 2.20	Short cell punctures
	pd2	Intracellular	1.25-3.71	3.07 ± 0.28	23.38 ± 2.70	
E1		Intracellular	0.49-2.05	1.08 ± 0.24	32.43 ± 1.80	Phloem salivation
E2		Intracellular	0.49-2.05	0.78 ± 0.20	34.53 ± 2.90	Phloem sap ingestion
G		Extracellular	1.37-3.00	1.86 ± 0.20	11.72 ± 0.30	Xylem sap ingestion

[Z] Relative amplitude (%) = (mean of amplitude for each waveform - mean of amplitude for non-probing)/ 5 ×100%

RESULTS AND DISCUSSION

Characterization of typical EPG waveforms for CMBS feeding behavior

EPG waveforms were characterized for CMBS when feeding on a host plant *L. limii* (Table 1), according to their shape, voltage level (extra- or intracellular), relative amplitude and frequency. These waveforms were labeled as C, pd1, pd2, E1, E2, and G.

Waveform C (Fig. 1A), correlating to gel salivation and other stylet pathway activities, was detected whenever CMBS started penetration and intercellular stylet pathway.

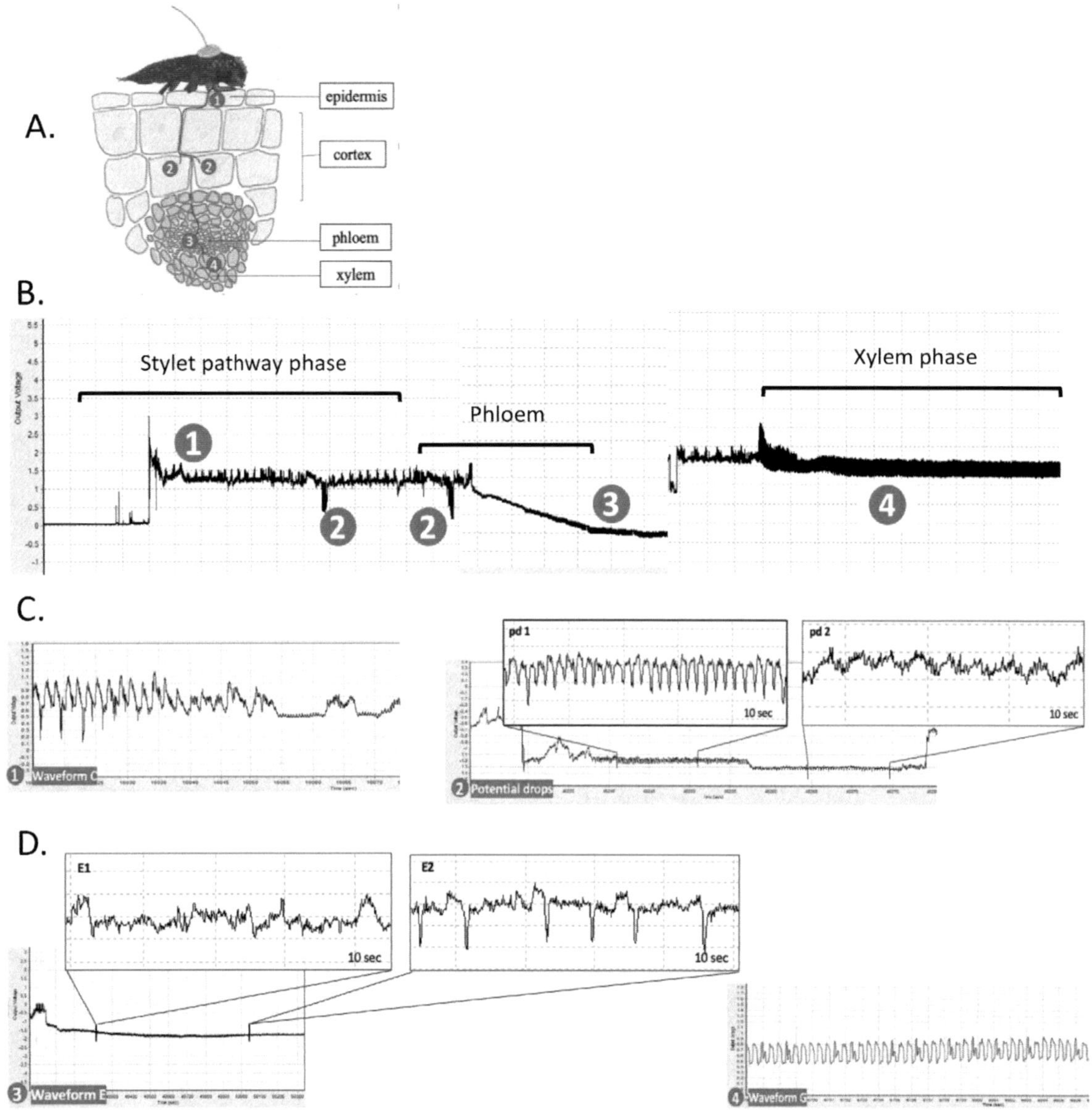

Figure 1 General scheme of characteristic feeding behavior of CMBS on *Lagerstroemia limii*. A: A diagram shows CMBS's stylet tip positions in a plant's stem when feeding. B: General scheme of characteristic EPG waveforms. C: ① Waveform C was detected when CMBS was probing intercellular part; ② Waveform potential drop (pd) was detected when the stylet tip punctured plant cells; ③ Waveform E1 was detected when intracellular stylet activity in mesophyll and phloem salivation occurred; Waveform E2, characterized by negative peaks, was detected when phloem sap ingestion occurred; ④ Waveform G was detected when xylem sap ingestion occurred.

Potential drops (Fig. 1B) were frequently observed during the stylet pathway phase. At the start, the voltage suddenly dropped when the stylet was supposed to puncture cells; during the low intracellular voltage level, the potential drops were often clearly divided into potential drop 1 (pd 1) and potential drop 2 (pd 2) periods. Waveform E complex (Fig. 1B), consisting of E1 and E2 phases during phloem phase, often sequentially followed the stylet pathway phase. The voltage level of the complex gradually and dramatically dropped below zero, which was much lower than other waveforms. Waveform E1, correlated to watery salivation, had positive peaks. Waveform G (Fig. 1C), correlated to xylem sap ingestion during the xylem phase, had a higher voltage level (extracellular) than other waveforms.

Comparison of feeding parameters of CMBS among different plant species

Even though plant species did not affect the total duration of waveform E1 ($F = 1.9326$; df = 7, 71; $P = 0.0769$), it affected the total duration of waveform C ($F = 6.8815$; df = 7, 71; $P < 0.0001$) and the total duration of waveform E2 ($F = 8.2204$; df = 7, 71; $P < 0.0001$) [Table 2]. After reaching the sieve elements, the insect spent the longest time in phloem sap ingestion on *L. limii* (234.78 ± 60.16 min) and the crapemrytle hybrid '18096' (286.43 ± 136.38 min), which was at least twice longer duration on *L. speciosa* (85.49 ± 38.84 min) and *C. acuminata* (19.84 ± 6.48 min). No individuals had phloem salivation or phloem ingestion on *F. pumila*, *F. auriculata*, or *G. max*.

Comparing with *F. auriculata*, even though the total duration of waveform G was multiple times greater on other species where the xylem ingestion occurred (varying from 90.27 ± 57.43 min to 423.54 ± 88.01 min), no significance was shown among the species [*L. limii*, *L. speciosa,* the hybrid '18096', *C. acuminata*, *F. tikuoa*, *F. pumila*, and *G. max* ($F = 1.8371$; df = 7, 71; $P = 0.0934$)].

From the perspective of stylet penetration activities, our study is the first report to elucidate the occurrences of phloem and xylem ingestion by CMBS on its host plant through the EPG techniques. We developed an R programming-based application to help identify and characterize the EPG waveforms with less human input. The comparison results of feeding parameters among different species indicated that CMBS accomplished the ingestion of phloem sap and xylem sap on the confirmed host plants (*C. acuminata*, *F. tikoua*, *L. limii*, *L. speciosa*) and the crapemyrtle hybrid '18096'. But CMBS did not intake phloem sap on *F. pumila*, *F. auriculata,* and *G. max*. With that, the "to eat, or not to eat" question was answered by applying the EPG techniques combined with computational analysis in the feeding behavior study of CMBS.

Supplementary

1) algorithm:
https://github.com/LylChun/epgminer

2) EPGminer in website version:
https://epgdata.shinyapps.io/epg-miner_app/

3) software version:
https://github.com/LylChun/epg-miner/tree/master/inst/epg-miner_app/rsconnect/shinyapps.io/epgdata

Table 2. Electrical penetration graph parameters of CMBS feeding on different plant species. C represents the stylet pathway phase; E1 represents phloem salivation; E2 represents phloem ingestion; G represents xylem ingestion.

Electrical penetration graph parameter	Plant Type			
	Lagerstroemia limii	*Lagerstroemia speciosa*	*Lagerstroemia indica* × *speciosa* **'18096'**	*Callicarpa acuminata*
Total duration of C (min)	488.65 ± 84.53 bc	428.32 ± 76.55 bc	671.19 ± 216.54 abc	549.78 ± 86.02 bc
Total duration of pd (min)	14.76 ± 3.25 ab	24.34 ± 7.00 ab	19.87 ± 4.54 ab	26.20 ± 9.25 ab
Total duration of E1 (min)	63.33 ± 32.46 a	35.37 ± 11.37 a	51.36 ± 19.15 a	47.90 ± 11.77 a
Total duration of E2 (min)	234.78 ± 60.16 a	85.49 ± 38.84 bc	286.43 ± 136.38 ab	19.84 ± 6.48 c
Total duration of G	288.52 ± 54.60 a	239.28 ± 86.96 a	90.27 ± 57.43 a	289.15 ± 64.38 a
Electrical penetration graph parameter	**Plant Type**			
	Ficus tikoua	*Ficus pumila*	*Ficus auriculata*	*Glycine max*
Total duration of C (min)	768.57 ± 54.24 ab	477.63 ±108.22 bc	1183.10 ± 153.45 a	262.20 ± 57.31 c
Total duration of pd (min)	43.90 ± 12.82 a	20.87 ± 7.23 ab	14.77 ± 11.76 ab	3.17 ± 1.74 b
Total duration of E1 (min)	39.64 ± 9.75 a	0.00 a	0.00 a	0.00 a
Total duration of E2 (min)	99.66 ± 26.79 abc	0.00 c	0.00 c	0.00 c
Total duration of G	317.46 ± 39.78 a	423.54 ± 88.01 a	0.00 a	190.79 ± 79.39 a

Means (± SE) followed by different letters within a row for each electrical penetration parameters were different by Tukey's Honestly Significant Difference (HSD) test (α =0.05).

The p-value for the comparison difference in the total duration of a certain waveform: C < 0.0001, pd = 0.0128, E1 = 0.0769, E2 <0.0001, and G = 0.0934.

LITERATURE CITED

EDDMapS. 2021. Early Detection & Distribution Mapping System. 15 March 2021. <https://www.eddmaps.org/distribution/viewmap.cfm?sub=80722>.

Gu, M., Merchant, M., Robbins, J. and Hopkins, J. (2014). Crape myrtle bark scale: A new exotic pest. EHT-049, 3. 11 May 2019. <https://cdn-ext.agnet.tamu.edu/wp-content/uploads/2018/10/EHT-049-crape-myrtle-bark-scale-a-new-exotic-pest.pdf>.

Hadley, W. (2016). Ggplot2: Elegrant graphics for data analysis. Springer.

Herbert, J.J., Mizell III, R.and McAuslane, H. (2009). Host preference of the crapemyrtle aphid (Hemiptera: Aphididae) and host suitability of crapemyrtle cultivars. Environ. Entomol. *38*, 1155-1160.

Kozár, F., Kaydan, M.B., Konczné, Z. Benedicty and Szita, .. (2013). Acanthococcidae and related families of the Palaearctic Region. Plant Protection Institute, Centre for Agricultural Research, Hungarian, Budapest, Hungary.

Merchant, M., Garcia, J. and Gu, M. (2018). Longevity of imidacloprid as a soil treatment for crapemyrtle bark scale on crapemyrtle, 2016–2017. Arthropod Management Test *43*:tsy012.

Prado, E. and Tjallingii ,W.F.. (1994). Aphid activities during sieve element punctures. Entomologia Experimentalis et Applicata *72*:157-165.

Prado, E. and Tjallingii, W.F. (1997). Effects of previous plant infestation on sieve element acceptance by two aphids. Entomologia Experimentalis et Applicata, *82*:189-200.

Schoonhoven, L.M., Van Loon, B., J.J. van Loon and Dicke, M. (2005). Insect-plant biology. Oxford University Press on Demand.

Sievert, C. (2020). Interactive web-based data visualization with R, plotly, and shiny. CRC Press.

Tjallingii, W. (1985). Electrical nature of recorded signals during stylet penetration by aphids. Entomologia Experimentalis et Applicata 38:177-186.

Tjallingii, W. and Esch, T.H. (1993). Fine structure of aphid stylet routes in plant tissues in correlation with EPG signals. Physiological Entomology *18*:317-328.

Tjallingii, W.F. (2006). Salivary secretions by aphids interacting with proteins of phloem wound responses. J. Experimental Botany *57*:739-745.

Wang, Z., Chen, Y. and Diaz, R. (2019). Temperature-dependent development and host range of crapemyrtle bark scale, *Acanthococcus lagerstroemiae* (Kuwana) (Hemiptera: Eriococcidae). Fla. Entomol. *102*:181-186.

Wu, B., Xie, R., Knox, G.W., Qin, H. and Gu, M. (2021). Host suitability for crapemyrtle bark scale (*Acanthococcus lagerstroemiae*) differed significantly among crapemyrtle species. Insects *12*:6.

Wu, X.H., Zhou, X.R. and Pang B.P. 2010. Influence of five host plants of *Aphis gossypii* Glover on some population parameters of *Hippodamia variegata* (Goeze). J. Pest Sci, *83*:7-83.

Xie, R., Wu, B., Dou, H., Liu, C., Knox, G.W., Qin, H. and Gu, M. (2020). Feeding preference of crapemyrtle bark scale (*Acanthococcus lagerstroemiae*) on different species. Insects *11*:399.

Zhang, Z. and Shi ,Y. (1986). Studies on the morphology and biology of *Eriococcus lagerstroemiae* kuwana. J. Anhui Agric. Univ., 2. 28 March 2020. <https://en.cnki.com.cn/Article_en/CJFDTotal-SCHO198602011.htm>.

Evaluation of Surfactants for Use in One-Time Foliar Auxin Applications in the Propagation of Woody Ornamentals

Anthony T. Bowden[a], Patricia R. Knight, Christine E.H. Coker, Jenny B. Ryals, Scott A. Langlois, Shaun R. Broderick, Eugene K. Blythe, Hamidou F. Sakhanokho, and Ebrahiem M. Babiker

Coastal Research and Extension Center, South Mississippi Branch Experiment Station, P.O. Box 193, Poplarville, MS 39470

ab1001@msstate.edu

[a]Second Place – Charlie Parkerson Graduate Student Research Paper Competition

Keywords: Adventitious rooting, indole-3-butyric acid (IBA), surfactant

Summary

Use of foliar applications are increasing in the nursery and greenhouse industries. However, previous research has shown that insufficient auxin is being absorbed or translocated to the site of action. Addition of surfactants to foliar applications of auxin may help with the absorption and translocation of auxin to the site of action. Research was conducted to determine whether addition of surfactants to one-time foliar applications of indole-3-butyric acid (IBA) would be as effective as the current industry standard, the basal quick dip. Terminal cuttings of common camellia (*Camellia japonica*) and Teddy Bear® magnolia (*Magnolia grandiflora* 'Southern Charm') were sprayed to the drip point using Hortus IBA Water Soluble Salts™ at concentrations of 0, 500, 1,000, or 1,500 ppm or dipped for 1-sec in a solution of either 4,000 or 2,500 ppm for camellia or magnolia, respectively. A foliar application of 1,500 ppm after sticking was as effective as the basal quick-dip for cuttings of Teddy Bear®, while other spray treatments were less effective. A basal quick-dip was more effective than a foliar spray for rooting cuttings of camellia.

INTRODUCTION

Research into foliar applications methods over the past decade indicated that one-time applications are the industry standard (Blythe et al., 2007; Kroin, 2014). When applied post-sticking, much lower concentrations (50 to 100 ppm) of rooting hormones are required compared to other conventional application methods (Dole and Gibson, 2006). Overhead applications of water soluble IBA are increasing in the nursery industry. Bailey Nurseries Inc. in Minnesota and Oregon has been conducting repetitive on-farm trialing for the last decade. Their results indicated that many of the taxa commonly propagated respond similarly to foliar-applied auxin compared to a traditional basal quick-dip. At Bailey Nurseries, propagation trays and beds are treated with a single application of water-soluble IBA ranging from 250 to 2,000 ppm (Drahn, 2007). Decker's Nursery in Ohio uses a battery-powered backpack sprayer to treat their cuttings since it atomized the auxin similarly to the mist from the mist system and applied a very small droplet with excellent coverage over both the top and bottom of the cutting (Decker, 2016). Since propagation areas vary in size, overhead applications are applied via a backpack sprayer for small houses and reel-and-hose sprayers for larger production areas.

When being applied overhead, Kroin (2014) of Hortus USA recommends to "spray the solution evenly over the cuttings until drops fall onto the media". To do this, Bailey Nurseries aims to deliver 1 L per 60 ft^2 (roughly 25-30 gal. per 6,000 ft^2). Currently, both Decker's Nursery and Bailey Nurseries generally treat their cuttings within 24-h of being stuck, either at the end of each day or the following morning, but application occurring during the day in conjunction with frequent mist intervals has not reduced efficacy (Drahn, 2007; Decker, 2016). Cuttings are treated in the early morning or late afternoon due to both lower light levels and reduced misting requirements. For both nurseries, the switch to overhead application led to a decrease in handling and the time cuttings spend in cold storage and the preparation room, where problems associated with lengthened exposure to low temperatures, high humidity, and/or handling can occur (Drahn, 2007). In 2003, 99.6% of cuttings at Bailey Nurseries were quick dipped and 0.4% were treated with foliar applications. By 2007, the percentages had reversed, with 95% of all propagated material being treated with overhead applications and 5.2% of material being quick-dipped (Drahn, 2007). Currently, overhead applications of water-soluble IBA are used to treat the following genera at Bailey Nurseries Minnesota operation: *Acer, Berberis, Cornus, Diervilla, Euonymus, Forsythia, Hydrangea, Juniperus, Lonicera, Philadelphus, Physocarpus, Rhus, Rosa, Spiraea, Symphoricarpos, Syringa, Thuja, Viburnum,* and *Weigela*. (Drahn, 2007).

Surfactants are common in agricultural production as penetration of the leaf cuticle is required for efficacy of foliar-applied compounds (Robertson and Kirkwood, 1969). Effectiveness of foliar-applied compounds depends on its ability to penetrate through the cuticle and translocate to the site of action (White et al., 2002). Surfactants enhance penetration of these chemicals by increasing the wetting capacity up to the critical micelle concentration (CMC), defined as the concentration above which any added surfactant molecules appear with high probability as micellar aggregates (Ruckenstein and Nagarajan, 1975; Lownds et al., 1987). Research was conducted by Lownds et al. (1987) to determine the effects surfactants would have on foliar penetration of NAA and NAA-induced ethylene production by cowpea [*Vigna unguiculata (*L.) Walp. subsp. *unguiculata* cv. Dixielee]. This research indicated that foliar penetration of NAA was increased when co-applied with a surfactant (Pace, Regulaid, or Tween 20) and all three induced similar qualitative changes in surface tension, contact angle, and droplet: leaf interaction. All three surfactants increased the droplet: leaf ratio. However, Regulaid was the only surfactant tested that showed a correlation between NAA penetration and interface area (Lownds et al., 1987). The objective of this research was to evaluate whether addition of surfactants to foliar auxin solutions increased root growth and uniformity compared to the industry-standard basal quick dip for common camellia (*Camellia japonica*) and Teddy Bear® magnolia (*Magnolia grandiflora* 'Southern Charm').

MATERIALS AND METHODS

Camellia

An experiment was performed to evaluate the effect of four foliar auxin concentrations [0, 500, 1,000, and 1,500 ppm IBA indole-3-butyric acid (IBA) (Hortus IBA Water Soluble Salts™; Phytotronics Inc., Earth City, MO)] each at two concentrations (0 and 0.85 ppm Regulaid®) on rooting of common camellia. Additionally, a basal quick-dip of 4,000 ppm IBA was used as an industry control. The experiment was a completely randomized design, consisting of an augmented factorial (4 auxin rates × 2 surfactant concentrations, plus the industry quick-dip control of 4000 ppm IBA). There were 15 cuttings per treatment, n=15.

Five-inch (12.7-cm), five-node terminal cuttings of *Camellia japonica* were harvested from established landscape plants and stuck to a depth of 0.5 inch (1.3 cm) on 18 August 2020. During cutting preparation, a one-inch (2.54 cm) wound was applied to one side of the basal end of cutting. The propagation medium was 100% pine bark placed into 3.5 inch (8.3 cm) square production pots (T.O. Plastics, Inc., Clearwater, MN). Cuttings receiving foliar applications of auxin were sprayed once to runoff with a 1-gal battery operated sprayer (One World Technologies, Inc., Anderson, SC). The pine bark propagation media was sourced from Eakes' Nursery Supply (Seminary, MS) and delivered as a mix of 50% aged and 50% fresh bark passed through a 3/8-inch (0.95 cm) screen. After treatment, cuttings were placed under intermittent mist applied for 6 sec/10 min during daylight hours and adjusted as needed for the studies duration.

The data was collected after 80 days included rooting percentage, shoot height, total root number, average root length (three longest roots), and root quality (1-5, with 1=no roots and 5= ≥ 10 roots). Additionally, net photosynthetic rate (A) and stomatal conductance (gs_w) values were sampled from five cuttings per treatment, for a total of 45 cuttings, between the hours of 7:30 A.M. and 11:30 A.M. using a LiCOR 6800 Portable Photosynthesis System (LI-COR Biosciences; Lincoln, NE). Data were analysed using linear mixed models and generalized linear mixed models with the GLIMMIX procedure of SAS (ver. 9.4; SAS Institute Inc., Cary, N.C.).

Teddy Bear® Magnolia

A similar experiment was done with Teddy Bear® magnolia which included four foliar auxin concentrations [0, 500, 1,000, and 1,500 ppm IBA indole-3-butyric acid (IBA) treated with two concentrations (0 or 0.85 ppm Regulaid®)]. However, the industry standard control was a quick-dip of 2500 ppm IBA. The experiment was also a completely randomized design, consisting of an augmented factorial (4 auxin rates × 2 surfactant concentrations, plus the industry quick-dip control of 2500 ppm IBA). There were 15 cuttings per treatment, n=15

Five-inch (12.7-cm), five-node terminal cuttings of *Magnolia grandiflora* 'Southern Charm' were harvested from established landscape plants and stuck to a depth of 0.5 inch (1.3 cm) on 14 April 2021. During cutting preparation, one-inch (2.54 cm.) wounds were applied to opposite sides of the basal end of the cutting. The propagation medium was 100% pine bark placed into 4.5 inch (12 cm) square production pots (T.O. Plastics, Inc., Clearwater, MN). Data collected after 125 days included rooting percentage, shoot height, total root number, average root length (three longest roots), and root quality (1-5, with 1=no roots and 5= ≥ 10 roots).

RESULTS

Camellia

Rooting percentage of camellia ranged from 23% to 50%, but surfactant nor auxin rate impacted rooting percentage (Table 1). Additionally, use of surfactant nor auxin rate had no effect on shoot height, or stomatal conductance (Table 1).

Table 1: Influence of surfactant and auxin rate on roots and shoots of a common camellia (*Camellia japonica*).

	Rooting (%)	Roots (no.)	Avg. length of three longest roots (cm)	Shoot height (cm)	Root quality rating[z]	Net photosynthesis ($\mu mol \cdot m^{-2} \cdot s^{-1}$)	Stomatal conductance ($mol \cdot m^{-2} \cdot s^{-1}$)
Surfactant:							
No surfactant	33	0.4b[y]	0.6b	0.3a	1.6b	2.3a	0.02a
Regulaid	50	1.4a	2.6a	0.2a	2.2a	1.4b	0.02a
Auxin Rate:							
0 ppm IBA	33	0.7bc	1.1a	0.3a	1.9a	2.0a	0.02a
500 ppm IBA	26	0.8b	0.4a	0.4a	2.0a	1.7a	0.02a
1,000 ppm IBA	23	0.5c	2.2a	0.2a	1.6a	1.7a	0.03a
1,500 ppm IBA	23	0.8bc	1.9a	0.4a	1.8a	1.9a	0.02a
4,000 ppm IBA	40	1.6a	2.5a	-	2.4a	1.8a	0.01a
Significance:							
Surfactant (S)	NS[x]	**	**	NS	**	**	NS
Auxin rate (A)	NS	**	NS	NS	NS	NS	NS
S × A	NS	NS	NS	NS	NS	NS	NS

[z]Root Quality (1-5, with 1 = no roots and 5 = ≥ 10 roots)
[y]Means within a column followed by the same letter were not different at $\alpha = 0.10$ or 0.05.
[x] Significant at the $P \leq 0.10$ (*) or 0.05 (**) level according to the Holm's Simulated Method.
NS= Not significant

Both surfactant and auxin had an effect on root number. Cuttings that were treated with 0.85 ppm Regulaid® had more roots compared to cuttings that received no surfactant (Table 1). Cuttings that were treated with 4,000 pm IBA as a quick-dip had more roots compared to cuttings that were treated with foliar applications of 0, 500, 1,000, or 1,5000 ppm IBA. The use of 0.85 ppm Regulaid® resulted in greater root lengths, greater root quality values, and lower net photosynthesis compared to cuttings not treated with surfactant at treatment initiation (Table 1).

Teddy Bear® Magnolia

Rooting percentage of Teddy Bear® magnolia ranged from 33% to 73% but neither surfactant nor auxin rate impacted rooting percentage (Table 2). Neither use of surfactant nor auxin rate had no effect on the average length of the three longest roots, net photosynthesis, or stomatal conductance values (Table 2). Treating cuttings with a foliar application of 1,500 ppm IBA or a 2,500 ppm IBA quick-dip - had more roots compared to cuttings that received a foliar auxin application of 0, 500, or 1,000 ppm IBA (Table 2).

Table 2: Influence of surfactant and auxin rate on roots and shoots of Teddy Bear® Southern magnolia (*Magnolia grandiflora* 'Southern Charm').

	Rooting (%)	Roots (no.)	Avg. length of three longest roots (cm)	Shoot height (cm)	Root quality rating[z]	Net photosynthesis ($\mu mol \cdot m^{-2} \cdot s^{-1}$)	Stomatal conductance ($mol \cdot m^{-2} \cdot s^{-1}$)
Surfactant:							
No surfactant	33	0.9a[y]	7.9a	0.6b	2.2b	6.3a	0.1a
Regulaid	73	1.2a	7.2a	1.3a	2.5a	6.3a	0.1a
Auxin Rate:							
0 ppm IBA	33	0.6b	6.4a	0.3b	1.8b	5.0a	0.1a
500 ppm IBA	33	0.3b	7.8a	0.5b	1.7b	5.9a	0.1a
1,000 ppm IBA	60	0.9b	9.1a	0.5b	2.1b	5.9a	0.1a
1,500 ppm IBA	53	1.6a	8.4a	1.2ab	2.8a	7.1a	0.1a
4,000 ppm IBA	60	1.8a	5.9a	2.4a	3.3a	7.5a	0.1a
Significance:							
Surfactant (S)	NS[x]	NS	NS	**	*	NS	NS
Auxin rate (A)	NS	**	NS	**	**	NS	NS
S × A	NS	NS	NS	NS	NS	NS	NS

[z]Root Quality (1-5, with 1 = no roots and 5 = ≥ 10 roots)
[y]Means within a column followed by the same letter were not different at $\alpha = 0.10$ or 0.05.
[x]Significant at the $P \leq 0.1$ (*) or 0.05 (**) level according to the Holm's Simulated Method.
NS= Not significant

Use of surfactant and auxin rate were found to be significant for shoot height and root quality (Table 2). Use of 0.85 ppm Regu-laid® resulted in greater shoot heights and root quality ratings than cuttings not receiving surfactant at treatment initiation. For shoot height, cuttings treated with a 2,500 ppm basal quick-dip resulted in a greater shoot length than cuttings treated with 0, 500 ppm, or 1,000 ppm IBA. Root quality rating for cuttings receiving foliar applications of 1,500 ppm or 2,500 pm IBA basal quick-dip were greater than cuttings treated with 0, 500 ppm, or 1,000 ppm IBA (Table 2).

DISCUSSION

Our results with *Camellia japonica* suggest that basal quick-dip is the preferred method for rooting cuttings of this species since other measured parameters were similar regardless of auxin treatment. However, if foliar applications of auxin are used, Regu-laid™ improved most results. Therefore, regardless of the use of surfactant, one-time foliar applications are not sufficient to improve rooting responses compared to the commercial standard basal quick-dip. Our results are similar to the propagation parameters for camellia (Dirr and Heuser, 2006; Dirr, 2009). Previous research into the propagation of camellia that auxin rates between 3,000 and 5,000 ppm stimulate adventitious rooting of camellia (Dirr and Heuser, 2006; Dirr, 2009).

The best rooting parameters for *M. grandiflora* 'Southern Charm' were obtained using a foliar spray of 1,500 ppm IBA or using a 2,500 ppm basal quick-dip compared to foliar applications of lower concentrations or untreated controls. One-time foliar applications of auxin appear to be of benefit for this species.

Our results from this trial and similar trials into foliar applications of auxin suggest that benefits of foliar applications are species dependent (Blythe et al., 2004). Our results suggest that sufficient auxin was absorbed from foliar applications and translocated to the site of root initiation so that root response is comparable to a basal quick-dip for Teddy Bear® magnolia but not for common camellia. By using a foliar application of 1,500 ppm IBA on a crop of Teddy Bear® magnolia, growers can eliminate the use of a basal quick-dip for propagation of this plant but using a foliar application of auxin on common camellias could result in up to 17% fewer rooted cuttings. With current methods, including quick-dips, propagators handle cuttings multiple times before flats enter the propagation house – resulting in higher labor costs for the producer. For crops where foliar auxin applications yield equal or better results compared to traditional quick-dips, propagators only need to handle the cutting once while a licensed applicator can treat cuttings using a backpack sprayer. In this manner, growers can increase profits by reducing labor costs. One grower reported a savings of $0.038/ft^2 (Drahn, 2007). In an industry where every penny affects the year-end profit, switching to foliar applications for 'Southern Charm' magnolia could potentially reduce labor costs and increase profits.

LITERATURE CITED

Blythe, E.K., Sibley, J.L., Ruter, J.M., and Tilt, K.M. (2004). Cutting propagation of foliage crops using a foliar application of auxin. Scientia Hort. *103*:31-37.

Blythe, E.K., Sibley, J.L., Tilt, K.M., and Ruter, J.M. (2007). Methods of auxin application in cutting propagation: A review of 70 years of scientific discovery and commercial practice. J. Environ. Hort. *25*:166-185.

Decker, B.M., and Graff, D. (2016). Foliar applications of rooting hormone at Decker Nursery. Comb. Proc. Intl. Plant Prop. Soc. *66*:191-196.

Dirr, M.A. (2009). Manual of woody landscape plants: Their identification, ornamental characteristics, culture, propagation, and uses. Stipes Publishing. Champaign, IL.

Dirr, M.A. and Heuser Jr., C.W. (2006). The reference manual of woody plant propagation: From seed to tissue culture-Second Edition. Varsity Press, Inc. Athens, GA.

Dole, J.M. and Gibson, J.L. (2006). Cutting Propagation: A guide to propagating and producing floriculture crops. Ball Publishing. Batavia, IL.

Drahn, S.R. (2007). Auxin application via foliar sprays. Comb. Proc. Intl. Plant Prop. Soc. *57*:274-277.

Kroin, J. (2014). Propagate plants from cuttings using foliar-applied aqueous (water-based) IBA rooting solutions: Tips-Do's and Don'ts. Comb. Proc. Intl. Plant Prop. Soc. *64*:149-159.

Lownds, N.K., Leon, J.M., and Bukovac M.J. (1987). Effect of surfactants on foliar penetration of NAA and NAA-induced ethylene evolution in cowpea. J. Amer. Soc. Hort. Sci. *112*:554-560.

Robertson, M.M. and Kirkwood, R.C. (1969). The mode of action of foliage-applied translocated herbicides with particular reference to the phenoxyacid compounds: 1. The mechanism and factors influencing herbicide absorption. Weed Res. *9*:224-240.

Ruckenstein, E. and Nagarajan, R. (1975). Critical micelle concentration: A transition point for micellar size distribution. J. Phys. Chem. *79*:2622-2626.

Growth of Liverwort (*Marchantia polymorpha*) and Spotted Spurge (*Euphorbia maculata*) Decreases with Substrate Stratification and Strategic Fertilizer Placement

Yuvraj Khamare[1a], S. Christopher Marble[2], James E Altland[3,] and Annette Chandler[4]

[1,2,4]University of Florida/IFAS, Mid-Florida Research and Education Center, Apopka, FL, USA, [3]Application Technology Research Unit, USDA-ARS, Wooster, OH, USA.

marblesc@ufl.edu

[a]Third Place – Charlie Parkerson Graduate Student Research Paper Competition

Keywords: Container media, resource efficiency, weed control, weed management

Summary

Substrate stratification is a method of filling nursery containers with "layers" of substrates (e.g., pine bark) comprised of different physical properties to manipulate soil moisture dynamics, improve irrigation and fertilization efficiency. However, stratification could also potentially serve as a weed management tool. The objective of this research was to assess the effect of stratified substrates and strategic fertilizer placement on the germination and growth of spotted spurge (*Euphorbia maculata*) and liverwort (*Marchantia polymorpha*) establishment in nursery pots. Before experiment initiation, aged pine bark was screened to three different sizes that consisted of particles ranging from 0.3 to 0.6 cm, 0.6 to 1.3 cm, and 1.3 to 1.9 cm. Bark was also screened to pass through a 1.3 cm and included all fines (all particle sizes less than 1.3 cm). The stratified treatments consisted of either the 0.3 to 0.6 cm, 0.6 to 1.3 cm, or 1.3 to 1.9 cm pine bark applied at depths of either 2.5 or 5 cm on top of the < 1.3 cm substrate. An industry-standard treatment was also included in which the substrate was not stratified but consisted of only the < 1.3 cm bark used throughout the container. A controlled-release fertilizer (CRF) was used at same rate

(35 g per pot) in all the treatments; However, fertilizer was incorporated in only the bottom layer in all stratified treatments (no fertilizer in the top 2.5 or 5 cm of the container media) while the industry standard had fertilizer incorporated throughout. Results showed that in comparison with the industry (non-stratified) standard, substrate stratification decreased spotted spurge germination by 30% to 84%. Spotted spurge shoot dry weight was reduced by 45% to 55% in stratified treatments when the top layer was applied at a depth of 5 cm, while a decrease of 14% to 42% was observed when the top layer was applied at a depth of 2.5 cm. Liverwort coverage was substantially reduced by nearly 100% in all the stratified substrate treatments. Overall, results suggest substrate stratification could be implemented in container production as part of an integrated weed management strategy.

INTRODUCTION

Weed management in container nurseries is currently primarily performed using frequent PRE herbicide applications in conjunction with hand weeding. High reliance on PRE herbicides causes several negative consequences such as high chemical costs, or concerns with recycling irrigation water (Poudyal and Cregg, 2019; Wilson et al., 1995). Additionally, the nursery industry produces thousands of different taxa and there is no herbicide labelled for use on all species. Popular plants such as succulents, herbaceous annuals, perennials, ornamental grasses, and tropical plants can also be highly sensitive to herbicides. When herbicides cannot be used, hand weeding costs can be significant. Darden and Neal (1999) reported that $1367 was spent to hand weed '1000' 3L (0.66 gal.) pots in just four months. There is a clear and immediate need to develop new, integrated and sustainable weed management strategies to reduce the cost of hand weeding and the disadvantages associated with a herbicide-only management strategy.

One cultural practice that has received some attention is the placement of controlled-release fertilizers (CRF) in pots. CRF is added to a nursery potting substrate to supply nutrients as the substrate used is mostly made up of materials such as bark, peat, perlite, or sand that lack nutrients. This ability to control nutrients in a pot can be strategically utilized to manage weed growth. Strategic fertilizer placement reduces weed growth by limiting the access to nutrients and at the same time increases the desired crop's competitive ability by direct access to nutrients (Nkebiwe et al., 2016). A fertilization method called subdressing has been shown to decrease the growth of spotted spurge (*Euphorbia maculata*) and eclipta (*Eclipta prostrata*) by over 80% (Saha et al., 2019; Stewart et al., 2018). Subdressing is accomplished by adding a layer of fertilizer in a pot filled with 50% to 75% potting substrate and filling the remainder of the pot with the same substrate (Khamare et al., 2020; Saha et al., 2019). This creates easy access to nutrients for crop roots, without any nutrients available for weed seedlings on the surface of the pot. Several studies have also shown that subdressing, can limit weed growth and reduce nutrient leaching without causing injury to the ornamental crops (Bir and Zondag, 1986; Stewart et al., 2018).

Another cultural method that could have potential for weed management is engineered substrates or substrate stratification. This is a new area of research that has the potential to decrease weed growth, water use, nutrient leaching that can result in reduced production time. Substrate stratification involves layering different substrates or the use of same substrate with different textures in a single pot (Fields et al., 2020). Fields et al (2021) reported that by using substrates with a high level of moisture and nutrient retention placed on top of a coarse, freely draining substrate, fertilizer rates could be reduced by 20% with no negative effects on the growth or quality of Red Drift roses (*Rosa* 'Meigalpio' PP17877) compared with an industry-standard substrate.

Theoretically, stratification of the top layer with freely draining, larger particle substrate without any fertilizer and the bottom layer with fine-textured, high moisture-retentive substrate could be used as a weed management tool. The top coarse-textured layer would hold less moisture and no fertilizer where weed seeds are introduced. Whereas the bottom layer would hold enough moisture and nutrients for the crop roots to access because as substrate particle size decreases, water holding capacity typically increases (Gruda and Schnitzler, 2004; Puustjarvi and Robertson, 1975; Richard and Beardsell, 1986). In this scenario, weed germination could be potentially reduced because weed seeds are introduced on the surface on the container substrate and require moisture for germination (Harper and Benton, 1996; Wada, 2005). Thus, it is possible that the top layer of substrate with less water holding capacity could result in reduced weed seed germination. Additionally, because the size of the most common container weed seeds is small, a top layer with a larger particle size could cause weed seeds to be flushed deep into the substrate, decreasing their chances of germination because many weed species require light to germinate (Keddy and Constabel, 1986).

Stratification could also potentially eliminate the disadvantages associated with mulching. First, the current industry practice is to fill the container with the same substrate with a space of 2 to 7 cm gap or more for mulch application (Altland et al., 2016; Bartley et al., 2017; Marble et al., 2019; Richardson et al., 2008) which reduces substrate volume and potential root growth. Mulching can also be costly, prone to blowing out of pots with high winds, or can be lost when pots are blown over. With substrate stratification, the extra step of mulching is eliminated as the top layer of stratified substrate will cover the pot surface and will be part of the growing substrate itself. As the plant liner is planted into this coarse bark layer, stratification would increase potential root volume compared with typical mulching practices. This would reduce the cost required for labor, mulching materials, and because crop roots would grow in this top stratified layer, less substrate would be lost due to wind or pot blow over. In theory, substrate stratification combines two of the most successful nonchemical weed management practices: strategic fertilizer placement and a 'mulch' like top layer that holds less moisture and no nutrients, but research is needed to verify these assumptions. The objective of this study was to evaluate the effect of substrate stratification on the growth of liverwort and spotted spurge.

MATERIALS AND METHODS

Experiments were conducted at the Mid-Florida Research and Education Center in Apopka, FL in 2020. Aged pine bark was purchased from a local supplier and further thoroughly screened to three different sizes that consisted of particles ranging from 0.3 to 0.6 cm, 0.6 to 1.3 cm, and 1.3 to 1.9 cm. An additional bark was also screened to pass through a 1.3 cm screen and included all fines (all particle sizes less than 1.3 cm). The stratified treatments were constructed by having either the 0.3 to 0.6 cm, 0.6 to 1.3 cm, or 1.3 to 1.9 cm bark as the top substrate with the bottom substrate consisting of ≤ 1.3 cm bark. The top substrate was applied at a depth of either 2.5 or 5 cm, resulting in six stratified substrate treatments (abbreviated as *top substrate size: screen size:" **S**" for stratification: top depth* in cm or 0.3-0.6:**S**:2.5, 0.3-0.6:**S**:5, 0.6-1.3:**S**:2.5, 0.6-1.3:**S**:5, 1.3-1.9:**S**:2.5 and 1.3 -1.9:**S**:5). An industry-standard treatment was also included in which the substrate was not stratified but consisted of only the ≤ 1.3 cm bark used throughout the container. A controlled-release fertilizer (CRF) (Osmocote® Blend 17-5-11 N-P-K [8 to 9 mo], ICL Specialty Fertilizers, Dublin, OH) at 35 g pot^{-1} was used at the same rate in all the treatments; However, fertilizer was incorporated in only the bottom layer in all stratified treatments (no fertilizer in the top 2.5 or 5 cm of the container media) while the industry standard treatment had fertilizer incorporated throughout. All the treatments consisted of pine bark and CRF without the addition of any other amendments such as peat moss or sand.

To assess weed growth, on Apr. 2020 and May. 2020, twenty-five seeds of spotted spurge were seeded in each pot to evaluate its growth and germination. Nursery pots (3.8 L) were filled and fertilized by the method mentioned above and seeds were surface sown. The pots were placed on a full sun nursery pad, irrigated 1.3 cm per day via overhead irrigation (Xcel® wobblers, Senninger Irrigation, Clermont, FL) via two irrigation cycles. Data collection included counts of emerged spotted spurge (mature and cotyledon) at 4 weeks after potting (WAP) and mature spotted spurge at 10 WAP. Shoot dry weight was collected at the trial conclusion (10 WAP). The experiment was a completely randomized design with eight single pot replication per treatment and repeated.

A separate set of nursery pots were used to evaluate liverwort (*Marchantia polymorpha*) growth on stratified substrates in Dec. 2020. Ten weeks before initiating the experiment and filling pots, 4 to 5 pieces of liverwort were transplanted onto the surface of 1.7 L nursery pots that had been previously filled with a pine bark: peat substrate (80:20 v: v) amended with the CRF via incorporation as described above. The pot was placed inside a shade house (60% ambient light) and was irrigated 1-cm per day via overhead irrigation. Pots remained in the shadehouse until the surface of the pots was filled with liverwort (no visible substrate upon visual inspection). At this time (approximately 10 wks after planting), these pots were used as inoculum to naturally sporulate the treatments as liverwort can spread asexually through the splashing of gemmae or sexually via airborne spores (Newby et al., 2007). Square 1.7 L nursery pots were filled and fertilized with the stratified and industry-standard treatments mentioned previously and placed inside the same shadehouse. To initiate the experiment, the inoculum pots were placed around each substrate treatment replication at a distance of 0.5 cm so that the experimental pots had an inoculum pot on all four sides. Liverwort surface coverage was assessed at 16 WAP by taking digital photos of each treatment using an iPhone (iPhone 8 Plus, Apple, Cupertino, CA) from a height of 0.9 m. Images were cropped using

Microsoft Paint (Microsoft Corp., Redmond, WA) so that only the surface of the substrate and liverwort was visible in the image. Liverwort coverage was then determined using the color threshold tool in ImageJ software (Abramoff et al., 2004). In all cases, data were subjected to analysis of variance using statistical software (JMP® Pro ver. 14, SAS Institute, Cary, NC). Prior to analysis, all data were inspected to ensure the assumptions of ANOVA were met. When appropriate, post hoc means comparisons were performed using Tukey's Honest Significant Differences test at a 0.05 significance level.

RESULTS AND DISCUSSION

Effect of substrate stratification on germination and growth of spotted spurge

At 4 WAP, spotted spurge germination was lower in most of the stratified substrate treatments in comparison with the 1.3:TO treatment (Table 1). The only exception was the 0.6-1.3:S:2.5 treatment which had similar germination in comparison with the 1.3:TO treatment. At 9 WAP, germination was still highest in the 1.3:TO treatment with reduced germination in all the stratified treatments. Overall, substrate stratification decreased spotted spurge germination by 30% to 84% in comparison with the industry-standard treatment of 1.3:TO (Table 1).

Shoot dry weight analysis showed that while germination was reduced, stratified treatments including 0.6-1.3:S:2.5 and 1.3-1.9:S:2.5 had shoot weight similar to the industry-standard treatment of 1.3:TO treatment. In the remaining stratified substrate treatments with a top layer of 5 cm (0.3-0.6:S:5, 0.6-1.3:S:5, 1.3-1.9:S:5), shoot weight decreased by 45% to 55% in comparison with the 1.3:TO treatment whereas shoot weight only decreased by 14% to 42% when the top layer was applied at a 2.5 cm depth.

Effect of substrate stratification on the establishment of liverwort

Liverwort growth was highest in the industry-standard treatment of 1.3:TO with an average coverage of 77% (Table 1). In all other treatments, liverwort coverage was negligible and less than 1% (Fig. 1). Liverwort thrives in an environment that has high moisture, high humidity, high fertility, and low ultraviolet light levels (Newby et al, 2007). As stratified substrates consist of a 2.5 to 5 cm of layer on top with low water holding capacity without any fertilizer, liverwort was unable to establish on the surface of the stratified treatments.

Overall, the growth of spotted spurge and liverwort was significantly reduced in the stratified substrates. Although not reported here for sake of brevity, additional experiments have been conducted with the same stratification technique described here with no adverse effects on some common ornamental species such as Japanese ligustrum (*Ligustrum japonicum*) and blue plumbago (*Plumbago auriculata*). Overall, current data suggest stratified substrates could be used as part of an overall integrated weed management program for container nurseries. Further research is ongoing to determine the impact of this method of substrate stratification on other weed and ornamental species.

Table 1. Effect of substrate composition on spotted spurge (*Euphorbia maculata*) germination and biomass and liverwort (*Marchantia polymorpha*) establishment.

	Spotted spurge			Liverwort
	Germination count[a]		Biomass[b]	% Coverage[d]
Substrate[c]	4WAP	9WAP	Shoot wt (g)	16WAP
1.3:TO	5.6 a[e]	11.4 a	22.4 a	77.2 a
0.3-0.6:S:2.5	1.6 c	5.1 bc	13.0 bc	0.4 b
0.3-0.6:S:5	0.9 c	3.5 c	10.0 c	0.3 b
0.6-1.3:S:2.5	3.9 ab	7.1 b	19.2 a	0.2 b
0.6-1.3:S:5	1.9 c	6.3 bc	12.4 c	0.02 b
1.3-1.9:S:2.5	2.4 bc	7.3 b	17.7 ab	0.02 b
1.3-1.9:S:5	0.9 c	4.6 bc	11.2 c	0 b

[a]Germination count was assessed by surface sowing 25 seeds of spotted spurge (*Euphorbia maculata*) to each pot and counting germinated seedlings at 4 weeks and 9 weeks after potting (WAP)

[b]Shoot dry wt was taken at trial conclusion at 10 weeks after seeding

[c]Substrate consisted of either the 0.3 to 0.6 cm, 0.6 to 1.3 cm, or 1.3 to 1.9 cm bark as the top substrate with the bottom substrate consisting of ≤ 1.3 cm bark and controlled release fertilizer (CRF) (Osmocote® Blend 17-5-11 N-P-K [8 to 9 mo]. The top substrate was applied at a depth of either 2.5 or 5 cm, resulting in six stratified substrate treatments (abbreviated as *top substrate size: screen size:" S" for stratification: top depth in cm* or 0.3-0.6:S:2.5, 0.3-0.6:S:5, 0.6-1.3:S:2.5, 0.6-1.3:S:5, 1.3-1.9:S:2.5 and 1.3 -1.9:S:5). An industry-standard treatment was also included in which the substrate was not stratified but consisted of only the ≤ 1.3 cm bark and CRF used throughout the pot

[d]Liverwort % coverage was measured by capturing photos at a height of 0.6 m above the pots and analyzed using the ImageJ software program at 16 WAP (week after potting) (5/22/2020)

[e]Means followed by the same letter within a column are not significantly different according to Tukey's HSD test $\alpha = 0.05$.

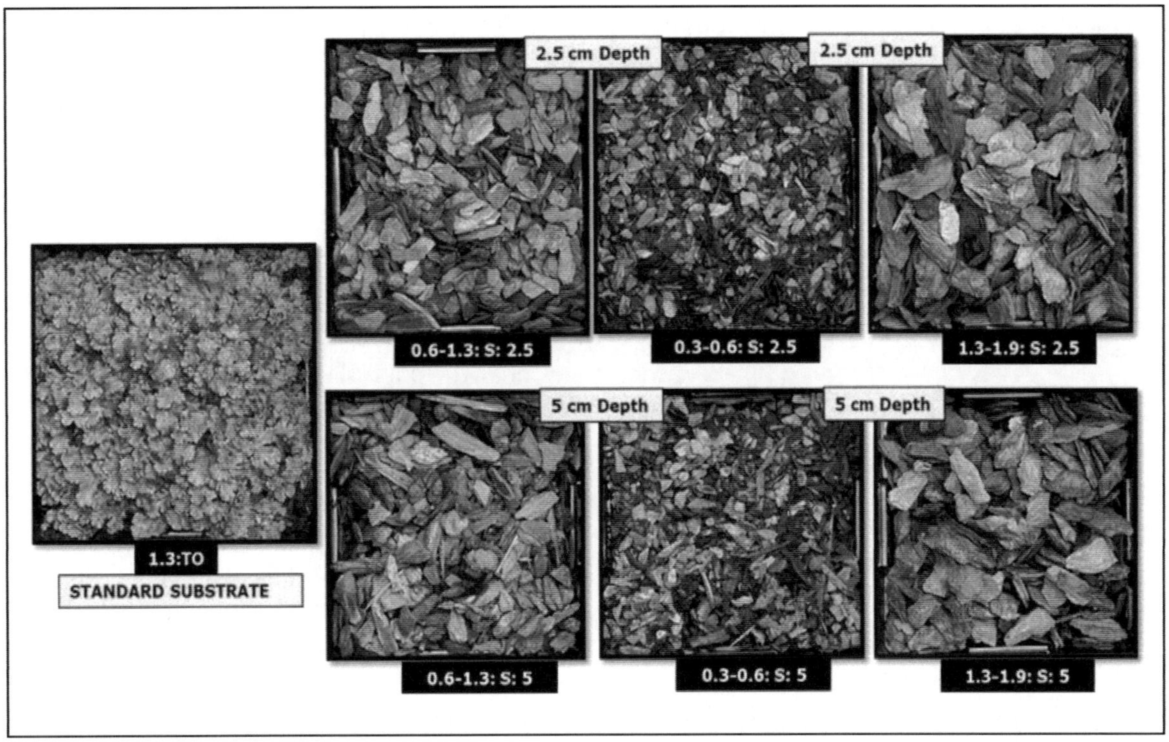

Figure 1. Liverwort (*Marchantia polymorpha*) coverage at 16 weeks after potting. Substrate consisted of either the 0.3 to 0.6 cm, 0.6 to 1.3 cm, or 1.3 to 1.9 cm bark as the top substrate with the bottom substrate consisting of ≤ 1.3 cm bark and controlled release fertilizer (CRF) (Osmocote® Blend 17-5-11 N-P-K [8 to 9 mo]. The top substrate was applied at a depth of either 2.5 or 5 cm, resulting in six stratified substrate treatments (abbreviated as top substrate size: screen size:" S" for stratification: top depth in cm or 0.3-0.6:S:2.5, 0.3-0.6:S:5, 0.6-1.3:S:2.5, 0.6-1.3:S:5, 1.3-1.9:S:2.5 and 1.3 -1.9:S:5). An industry-standard treatment was also included in which the substrate was not stratified but consisted of only the ≤ 1.3 cm bark and CRF used throughout the pot.

LITERATURE CITED

Abramoff, M., Magalhaes, P., and Ram, S. (2004). Image processing with ImageJ. Biophotonics International. Laurin Publishing. < http://dspace.library.uu.nl/handle/1874/204900>

Altland, J.E., Boldt, J.K., and Krause, C.C. (2016). Rice hull mulch affects germination of bittercress and creeping woodsorrel in container plant culture. Amer. J. Plant Sci. *7*:2359-2375.

Bartley, P.C., Wehtje, G.R., Murphy, A.M., Foshee, W.G., and Gilliam, C.H. (2017). Mulch type and depth influences control of three major weed species in nursery container production. HortTechnology *27*:465-471.

Bir, R.E. and Zondag, R.H. 1986. The great dibble debate: Test results raise more question. Amer. Nurse. *164*:59-64.

Darden, J. and Neal, J.C. (1999). Granular herbicide application uniformity and efficacy in container nurseries. Proc. Southern Nursery Assn Res. Conf. *44*:427–430.

Fields, J.S., Owen, J.S., and Altland J.E. (2020). Stratified substrates: A media management strategy for increased resource efficiency. HortScience, *55*:S399–S400.

Fields, J.S., J.S. Owen, J.E. Altland. 2021. Substrate Stratification: layering unique substrates within a container increases resource efficiency without impacting growth of shrub rose. Agronomy. *11*:1454.

Gruda, N., and Schnitzler, W.H. (2004). Suitability of wood fiber substrate for production of vegetable transplants. I. Physical properties of wood fiber substrates Scientia Hort. *100*:309-322.

Harper, J., and Benton, R. (1966). The behavior of seeds in soil. II. The germination of seeds on the surface of a water supplying substrate. J. Ecol. *54*:151-166.

Keddy, P.A., and Constabel, P. (1986). Germination of ten shoreline plants in relation to seed size, soil particle size and water level: an experimental study. J. Ecol. *74*:133-141.

Khamare, Y., Marble, S.C., and Chandler, A. (2020). Fertilizer placement effects on eclipta (*Eclipta prostrata*) growth and competition with container-grown ornamentals. Weed Sci. *68*:496-502.

Marble, S.C., Steed, S.T., Saha, D., and Khamare, Y. (2019). On-farm evaluations of wood-derived, waste paper, and plastic mulch materials for weed control in Florida container nurseries. HortTechnology *29*:866-873.

Newby, A., Altland, J.E., Gilliam, C.H., and Wehtje, G. (2007). Pre-emergence liverwort control in nursery containers, HortTechnology *17*:496-500.

Nkebiwe, P.M., Weinmann, M., Bar-Tal, A., Muller, T. (2016). Fertilizer placement to improve crop nutrient acquisition and yield: a review and meta-analysis. Field Crops Res. *196*:389–401.

Poudyal, S. and Cregg, B.M. (2019). Irrigation nursery crops with recycled run-off: a review of potential impact of pesticides on plant growth and physiology. HortTechnology *29*:716-729.

Puustjarvi, V., and Robertson, R.A. (1975). Physical and chemical properties. p 23–38. In: D.W. Robinson and J.G.D. Lamb (eds.). Peat in Horticulture. Academic Press, London.

Richards, D., Lane, M. and Beardsell, D.V. (1986). The influence of particle-size distribution in pinebark:sand:brown coal potting mixes on water supply, aeration, and plant growth. Scientia Hort. *29*:1-14.

Richardson, B., Gilliam, C.H., Fain, G., and Wehtje, G. (2008). Nursery container weed control with pinebark mininuggets. J. Environ. Hort. *26*:144-148.

Saha, D., Marble, S.C., Torres, N., and Chandler, A. (2019). Fertilizer placement affects growth and reproduction of three common weed species in pine bark-based soilless nursery substrates. Weed Sci. *67*:682-688.

Stewart, C., Marble, S.C., Jackson, B.E., Pearson, B.J., and Wilson, P.C. (2018). Effects of three fertilization methods on weed growth and herbicide performance in soilless nursery substrates. J. Environ. Hort. *36*:133-139.

Wada, S. (2005). Nursery container weeds response to modification of substrate pH, substrate particle size and applied nitrogen form. Oregon State Univ., Corvallis, MS Thesis, 72–76.

Wilson, P.C., Whitwell, T., and Riley, M.B. (1995). Effects of ground cover and formulation on herbicides in runoff water from miniature nursery sites. Weed Sci. *43*:671-677.

Exploring Water Movement through Stratified Substrates

Kristopher S. Criscione[1], Jeb S. Fields[1], and James S. Owen Jr.[2]

[1] LSU AgCenter Hammond Research Station, 21549 Old Covington Highway, Hammond, Louisiana 70403; [2] United States Department of Agriculture Application Technology Research Unit, 1680 Madison Ave., Wooster, OH 44691

JFields@agcenter.lsu.edu

Keywords: Container media, container water dynamics, irrigation, water efficiency

Summary

An increase in horticultural production requires a greater demand for more water use. Soilless substrates, particularly bark-based systems used in nursery production, can be inefficient with regards to resource utilization. Substrate stratification is an innovative substrate management technique that involves the layering or stacking two substrates of unique hydraulics properties within the container system. The objective of this study was to monitor how stratifying substrates influences substrate water potential between two different irrigation schedules. Stratified substrates allow for added water retention in the upper half of the container, whereas in the lower half, air-filled porosity was increased. Moreover, stratified substrates significantly reduced tension fluctuations that notoriously occur in the upper portion of the substrate profile. Oscillations were even further reduced when a cyclic irrigation schedule was implemented. Thus, stratified substrates have potential for improving water efficiency in nursery crop production.

INTRODUCTION

Agricultural production continues to be a primary consumer for natural resource withdrawals, specifically water, in the United States (Calzadilla et al., 2010). The nursery industry is a growing agricultural sector, where annual sales have increased $3 billion within the last decade according to the Census of Agriculture (USDA, 2019). Water scrutiny, availability, and local regulatory restrictions acknowledge the current challenges the nursery industry faces, especially with regards to horticultural substrates (Fulcher et al., 2016).

Nursery crops are conventionally produced in containers filled uniformly with a singular or multicomponent substrate, typically bark-based, and utilized for their suitable drainage and aeration properties (Pokorny, 1979). However, soilless substrates are inefficient in regard to resource use (water and mineral nutrients), requiring daily irrigation applications and continuous fertilization (Tyler et al., 1996). The need for constant irrigation is due to the limited container volume and the high porosity of bark-based substrates, which creates an undesirable moisture gradient (i.e. the upper portion of the pot is drier than the lower portion). Thus, resulting in an increase in water use to replenish the finite amount of available water (Owen and Altland, 2008). Therefore, engineering horticultural substrates to control water gradients within the container may result in more resource efficient production practices.

Substrate stratification, layering of unique substrates within the container to modify the air to water ratio for more desirable water retention and drainage properties, is a substrate management strategy that may improve plant nursery resource efficiency (Fields et al., 2021). Layering fine or fibrous substrate particles in the upper half of the container may increase substrate water holding capabilities in the initial plug or liner growing area, whereas the arrangement of coarse particles in the lower half can increase aeration and substrate drainage. This ability to engineer the hydraulic gradient within a container may be further benefited from precision irrigation scheduling, wherein water can be applied to supply the upper portion of the container.

Water availability which is associated with plant stress, quality, yield, and subsequent abilities for root systems to overcome drying periods can be estimated as a substrate tension (Shock et al., 2011). Substrate tension is a measure of how tightly water is held within a substrate and is commonly measured through use of tensiometers. A substrate that is able to withstand reaching low tensions would ensure plant roots can readily access water and nutrients. This in turn could not only improve crop quality, but may also lead to improved resource efficiency. Wallach (2008) discussed the use of tensiometers in the top and bottom portion of a nursery container filled with perlite under 'moist' and 'dry' conditions. It was observed that more frequent irrigations (moist) increased substrate water potentials (less negative) in the upper half of the substrate profile. Thus, improving water holding in the upper 50% of the container profile by placement of fine particles should further increase water potentials for more desirable substrate tensions.

Therefore, the objective of this study was to monitor and compare substrate water potentials throughout the container system of non-stratified and stratified profiles during daily water fluctuations and draw comparisons between two irrigation schedules (single application or cyclic).

MATERIALS AND METHODS

Aged loblolly pine (*Pinus taeda* L.) bark particles were fractioned via a continuous flow screen utilizing a 6.3 mm aperture screen. Conventional bark (unscreened), fine bark particles (< 6.3mm), and coarse bark particles (> 6.3 mm) were collected. Substrate physical properties were assessed on three replicates of each substrate utilizing a NCSU porometer to measure container capacity (CC), air space (AS), total porosity (TP) and bulk density (D_b; Fonteno and Harden, 2010). Substrate particle size distribution was measured by passing three 100 g dry replicates of each bark through a series of sieves while agitating for five min with a screen shaker (Ro-Tap Shaker; W.S. Tyler, Mentor, OH) and weighing the particles remaining on each screen. Substrate hydraulic properties were also assessed on three replicates of each material utilizing the evaporative method described Fields et al. (2016).

Twenty containers (5.68 L) were filled with either of two substrates treatments 1) a conventional bark substrate or 2) a stratified substrate where coarse bark was utilized to fill to lower half the container and fine bark was utilized to fill the upper half of the container. Six replicates of each substrate treatment were fitted with calibrated elbow tensiometers (Soil Measurement Systems; Huntington Beach, CA, USA) at 25% and 75% below the substrate surface (Fig. 1). The replicates were randomly split into two irrigation treatments in a climate-controlled greenhouse. Irrigation treatments consisted of a single application irrigation schedule (SI; 1x/d, 600 mL) and cyclic application irrigation schedule (CI; 3x/d, 200 mL; 600 mL total). Data was collected for 6 d in fallow pots and recorded with a data logger (CR1000X; Campbell Scientific, Logan UT, US). Data was analysed using JMP Pro (15.1.0; SAS Institute, Inc.; Cary, NC, U.S.) utilizing Tukey's Honestly Significant Difference ($\alpha = 0.05$) to separate means across substrates (Table 1).

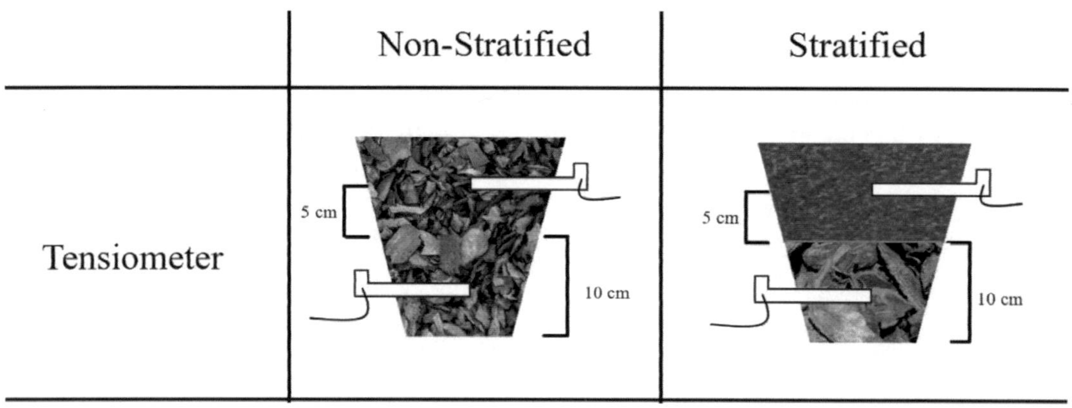

Figure 1. Fictitious depiction of tensiometer installation dimensions and placement.

RESULTS AND DISCUSSION

Physical Properties

Partitioning bark particles smaller than 6.3 mm significantly increased the substrate's ability to retain water (0.52 $cm^3 cm^{-3}$) while increasing the majority of bark particle diameter greater than 6.3 mm reduced substrate CC (0.39 $cm^3 cm^{-3}$), when compared to conventional bark (0.46 $cm^3 cm^{-3}$; Table 1). Bilderback et al. (2013) suggest that container capacity for horticultural substrates should range from 0.45-0.65 $cm^3 cm^{-3}$. Increasing particle diameter from conventional bark also increases AS; however, reducing particle diameter did not influence AS (Table 1).

Table 1. Static physical properties and particle size distribution of pine bark substrates utilized in stratified substrate systems. Conventional bark was fractioned by passing through a 6.3 mm screen. The particles that remained on the screen were considered coarse bark, and the particles that passed through the screen were considered fine bark.

	Static Physical Properties [a]			
Substrate	Container capacity $cm^3 cm^{-3}$	Air space $cm^3 cm^{-3}$	Total porosity $cm^3 cm^{-3}$	Bulk density $g cm^{-3}$
Conventional bark	0.46 b [c]	0.33 b	0.79 a	0.17 a
Fine bark	0.52 a	0.30 b	0.82 a	0.17 a
Coarse bark	0.39 c	0.43 a	0.83 a	0.16 a
P-value [d]	<0.0001	0.0098	0.4252	0.0956

	Particle Size Distribution[b]			
	Extra Large (>6.3 mm) $g g^{-1}$	Large (6.3–2.00 mm) $g g^{-1}$	Medium (2.00-0.71 mm) $g g^{-1}$	Fines (<0.71 mm) $g g^{-1}$
Conventional bark	0.36 b	0.43 b	0.13 b	0.08 b
Fine bark	0.01 c	0.50 a	0.36 a	0.14 a
Coarse bark	0.56 a	0.36 c	0.04 c	0.04 c
P-value [d]	<0.0001	<0.00021	0.0001	0.0019

[a] Measured via porometer analysis. Total porosity = air space (minimum air-filled porosity after free drainage) + container capacity (maximum water holding capacity after free drainage).
[b] Percent of total sample dry mass within the particle size range.
[c] Letters denote detected differences among means of three substrates (conventional bark, fine bark, and coarse bark) utilizing Tukey's HSD ($\alpha = 0.05$).
[d] Measures of overall treatment effects utilizing ANOVA analysis with a significance value of ($\alpha = 0.05$).

Again, coarse bark was the only substrate that was not within recommended guidelines for CC and AS (0.10-0.30 $cm^3\ cm^{-3}$; Bilderback et al., 2013). Total porosity and bulk density were unaffected by fractionating bark particles (Table 1).

Concentrating the majority of bark particles greater than 6.3 mm resulted in the greatest proportion of extra-large particles and alternatively, reducing particle size significantly decreased the percentage of extra-large particles relative to conventional bark (Table 1). Inversely, fine bark particles had the greatest proportions of large, medium, and fine particles (i.e. <6.3 mm), whereas coarse particles had the least (Table 1).

Hydraulic Properties

Substrate hydraulic properties were utilized to develop moisture characteristic curves, which were subsequently fit to a constrained soil water retention model (van Genuchten, 1985). The porosity of conventional pine bark is heterogeneous, which results in a non-uniform pore size distribution. Thus, a myriad of pore sizes exists throughout a nursery container filled with conventional bark (Drzal et al., 1999). However, the bark screening process creates a more uniform pore size distribution due to the bulk of the bark particle sizes consisting of semi-identical diameters (Fields et al., 2018). A gradual decline in volumetric water content (VWC) with decreasing tension was observed in conventional bark (Fig. 2A), which confirms heterogeneous porosity. Moreover, the conventional bark retains more water at lower tensions than the other substrates do (Fig. 2). This is likely due to water being restrictively held through hysteretic porosity throughout the profile. Conversely, fine and coarse bark have a rapid decline in VWC below tensions considered readily available water (-10 and -50 hPa; de Boodt and Verdonck, 1972), likely due to the uniform pore size distribution, indicative of the screening process (Fields et al., 2021; Fig. 2B).

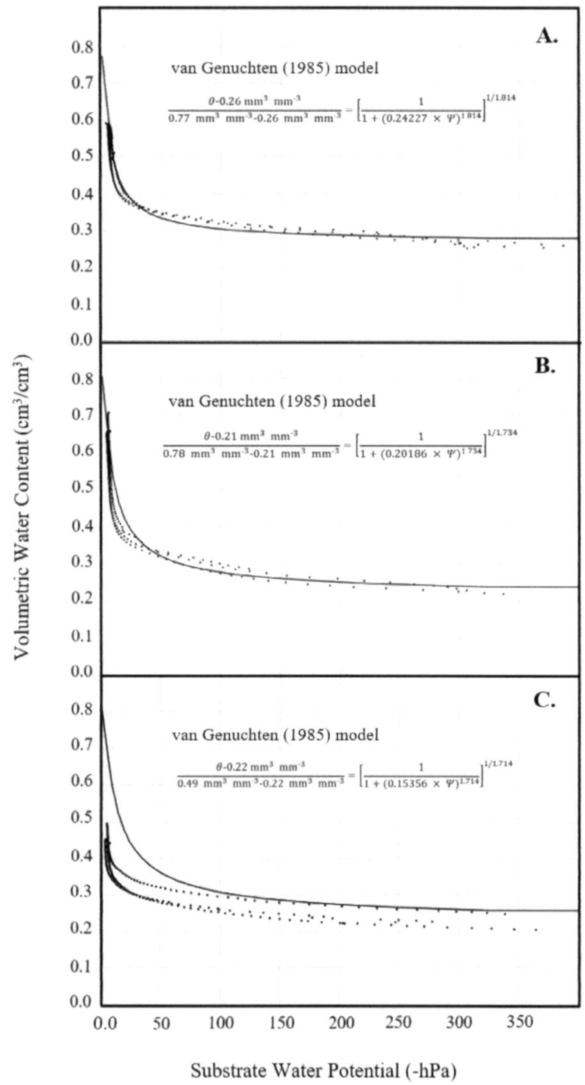

Figure 2. Substrate moisture characteristic data (points) fit to a constrained van Genuchten (1985) hydraulic model (solid line). The data was measured via evaporative measurement and porometers on three replicates of each substrate. Substrates include A) conventional bark, B) fine bark (<6.3 mm) and C) coarse bark (>6.3 mm). Volumetric water content ($cm^3\ cm^{-3}$; Y-axis) was plotted against substrate water potential (-hPa; X-axis).

The fine bark also had the greatest initial CC when compared to other substrates (Table 1). The coarse bark had greater particle diameters, resulting in an increased percentage of macropores (Drzal et al., 1999). Thus, the rate of water loss in the coarse bark diminished at relatively high tensions as there was little remaining free water, where small reductions in VWC continued to result in large reductions in water potential in at higher tensions than the other barks leaving the remainder of water tightly surface bound (Fig. 2C).

Monitoring Substrate Water Potential Non-stratified substrates experienced large fluctuations in daily substrate tensions when receiving a single irrigation event (Fig. 3A).

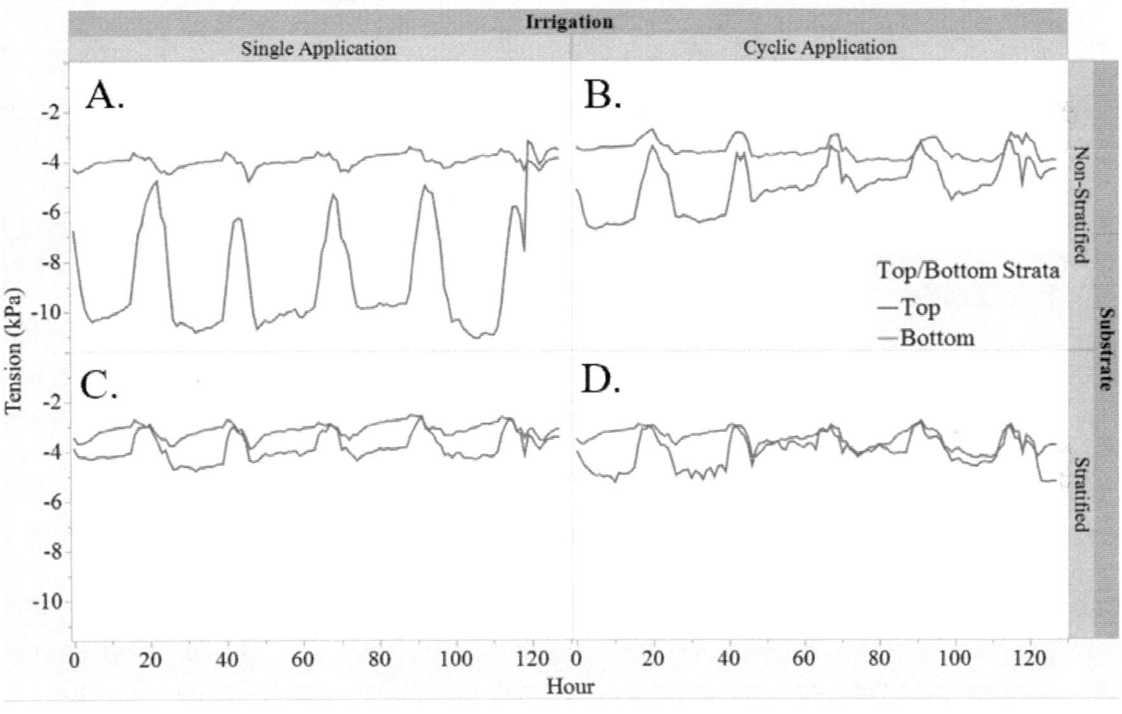

Figure 3. Substrate water potentials calculated via elbow tensiometers in the upper and lower portions of the container under single or cyclic irrigation application scheduling over 6 d. Treatments include A) Non-stratified substrates within a single application irrigation B) Non-stratified substrates within a cyclic irrigation application C) Stratified substrates in a single application and D) Stratified substrates within a cyclic irrigation application.

The lowest tension reached (below -10 kPa) was in the upper portion of the non-stratified substrate under a single application. This is indicative of a traditional container substrate system, where the upper proportion of the substrate dries rapidly due to gravitational drainage and evaporation (Fonteno, 1989). The stratified system reached tensions only half of that in the same irrigation schedule (-5 kPa; Fig. 3C). Stratifying the substrate reduced the water loss from gravitational drainage through increased upper strata water retention, maintaining a more continual moisture profile, while the greater proportion of extra-large particles in the coarse bark resulted in rapid water loss in the lower 50% of the substrate profile (Fig. 3A). Hillel (2004) stated that VWC and tension are inversely related; hence, more water holding capabilities ensured tensions remained within the range of readily available water, possibly reducing energy required for root uptake (Fields, 2016).

Cyclic application scheduling effectively increased substrate tensions to more desirable water potentials in the control substrate (Fig. 3B). The more frequent and shallow irrigations increased the VWC in the upper half of the container for longer durations where it was observed to have dried to a greater magnitude in a single, large irrigation application (Fig. 3A-B). To a greater extent, when stratified substrates consist of fine bark particles on the top, an optimal substrate tension was maintained throughout the day (Fig. 3D). Moreover, in the stratified system under cyclic irrigation, the tension in the upper half followed parallel trends with the tension in the lower half during and between irrigation events. This is evidence of the uniform water gradient within the container system that was hypothesized to result from the stratifying process (Fig. 3D).

Through most of the monitoring, the upper portion of the container experienced the greatest daily fluctuation in water potential. All lower strata water potentials were relatively stable with minimal deviations (± -2 kPa; Fig. 3). This indicates that incorporation of coarse bark materials in the lower portion of the container system did not adversely affect moisture content, instead they provided relatively stable water potentials through production. Thus, stratifying substrates were able to optimize upper container water balance where the initial plant rooting zone occurs (i.e. from initial liner or plug growth) while maintaining optimal lower container VWC.

It is important to develop and engineer more resource efficient production practices as the horticultural industry continues to increase in production. Stratifying the substrate through layering fine bark on top of coarse bark has been shown as a method to effectively reduce daily water fluctuations within the container while maintaining optimal water tensions throughout the container system. Furthermore, pairing stratified substrates with more efficient and targeted irrigation strategies (i.e. cyclic irrigation) can further stabilize substrate moisture tensions during and between irrigation events. Traditional nursery substrates irrigated daily will experience large changes in in water potential in the profile. Stratified substrates greatly reduce the tension fluctuations through strategic modified substrate hydraulic modifications.

LITERATURE CITED

Altland, J. E., Owen, Jr., J.S., Jackson, B.E. and Fields, J.S. (2018). Physical and hydraulic properties of commercial pine-bark substrate products used in production of containerized crops. Hort. Sci. *53*:1883-1890.

Bilderback, T.E., Owen, Jr., J.S., Altland, J.E., Fain, G.B., Jackson, B.E., Riley, E.D., Kraus, H.T. and Fonteno, W.C. (2013). Strategies for developing sustainable substrates in nursery crop production. Acta. Hort. *1013*:43-56.

Calzadilla, A., Rehdanz, K. and Tol, R.S.J. (2010). The economic impact of more sustainable water use in agriculture: A computable general equilibrium analysis. J. Hydro. *384*:292-305.

de Boodt, M. and Verdonck, O. (1972). The physical properties of substrates in horticulture. Acta Hort. *26*:37-44.

Drzal, M.S., Fonteno, W.C. and Cassel, K.C. (1999). Pore fraction analysis: a new tool for substrate testing. Acta Hort. *481*:43-54.

Fields, J.S., Owen, Jr., J.S., Zhang, L. and Fonteno, W.C. (2016). The use of the evaporative method for determination of soilless substrate moisture characteristic curves. Scientia Hort. *211*:102-109.

Fields, J.S., Owen, Jr., J.S., Altland, J.E., van Iersel, M.W. and Jackson, B.E. (2018). Soilless substrate hydrology can be engineered to influence plant water status for an ornamental containerized crop grown within optimal water potentials. J. Amer. Soc. Hort. Sci. *143*:268-281.

Fields, J.S., Owen, Jr., J.S. and Altland, J.E. (2021). Substrate Stratification: Layering Unique Substrates within a Container Increases Resource Efficiency without Impacting Growth of Shrub Rose. Agron. J. *11*:1454.

Fonteno, W.C. (1989). An approach to modeling air and water status of horticultural substrates. Acta Hort. *238*:67-74.

Fonteno, W.C. and Harden, C.T. (2010). North Carolina State University Horticultural Substrates Lab Manual. North Carolina State University.

Fulcher, A., LeBude, A.V., Owen, Jr., J.S., White, S.A. and Beeson, R.C. (2016). The next ten years: strategic vision of water resources for nursery producers. Hort. Tech. *26*:121–132.

Hillel, D. (2004). Introduction to environmental soil physics. Elsevier Academic Press, San Deigo, CA.

Pokorny, F.A. (1979). Pine bark container media - an overview. Proc. Symp. Inter. Plant Prop. Soc. *29*:484-494.

Shock, C.C., and Wang, F. (2011). Soil water tension, a powerful measurement for productivity and stewardship. Hort. Sci. *46:*178-185.

Tyler, H.H., Warren, S.L. and Bilderback, T.E. (1996). Cyclic irrigation increases irrigation application efficiency and decreases ammonium losses. J. Environ. Hort. *14*:194-198.

U.S. Department of Agriculture: (2019). 2017 Census of Agriculture. In N. A. S. Service (Ed.), USDA, p. 52 p.

van Genuchten, M.T. and Neilson, D.R. (1985). On describing and predicting the hydraulic properties of unsaturated soils. Annales Geophysicae *3*:615-628.

Wallach, R. (2008). Physical characteristics of soilless media, Elsevier, San Diego, CA.

Rooting Response of Piedmont Azalea (*Rhododendron canescens)* Softwood, Single-node Cuttings to a Basal Auxin Quick Dip

Jenny B. Ryals, Patricia R. Knight, Christine E. H. Coker, Gary R. Bachman, Jim M. DelPrince, Patricia R. Drackett, Scott Langlois, David Brand and Anthony T. Bowden

Coastal Research and Extension Center, South Mississippi Branch Experiment Station P.O. Box 193, Poplarville, Mississippi 39470

j.ryals@msstate.edu

Keywords: Propagation, IBA, Hortus IBA water soluble salts

Summary

Piedmont azalea (*Rhododendron canescens)* is a deciduous azalea native to the southeastern United States as well as areas in Maryland and Pennsylvania. Cutting propagation reduces the variability observed when propagating from seed. As a whole, deciduous azaleas are known to be difficult to root via cuttings, however, piedmont azalea has been reported as moderate to easy to propagate from softwood cuttings. Piedmont azalea has been observed to root as softwood cuttings treated with a range of indole-3-butyric acid (IBA) quick dips from 5,000 to 10,000 ppm. The objective of this research was to determine rooting response of very soft single node cuttings to a basal auxin quick dip in order to provide growers with relevant cutting propagation recommendations. Naturally occurring auxins are produced in newly forming tissues. Therefore, a low dose of endogenous auxin might encourage young plants to root faster and more efficiently than older cuttings. Results indicate that single node Piedmont azalea cuttings will root with or without the use of an auxin basal quick dip.

INTRODUCTION

Propagation of native deciduous azaleas (*Rhododendron* spp.) can be done by seed, cutting, and layering. Deciduous azaleas are considered to be a difficult to root plant species. Due to trait variability observed in seed-grown azaleas, cutting propagation is preferred; however, cutting propagation recommendations can vary (Hyatt, 2006; Sommerville, 1998). According to Dirr and Heuser (2018), slightly firm, 15.2 cm (6 in.) cuttings should be taken from the beginning to the end of April. Dirr and Heuser (2018) also recommend using a fungicide with 4000 ppm IBA dip; however, recommended auxin concentrations can vary with different cultivars. Hyatt (2006) recommends taking 5 to 8 cm (2 to 3 in.) softwood cuttings in late May to early June while the plants are actively growing. Bir (1992) achieved successful rooting with softwood cuttings taken after new growth had ceased and treated with 1000-2500 ppm IBA. Ryals et al. (2019) achieved low percentages of rooting when cuttings of new plant tissue were taken after flower senescence and treated with 2500 ppm IBA.

Besides being difficult to root, deciduous azaleas can also be problematic to break dormancy and put on new growth (Brown, 2017). According to Hyatt (2006), the stronger the rooting hormone used, the more difficult it is for cuttings to break dormancy and start actively growing. Piedmont azalea, (*Rhododendron canescens* [Michaux.] Sweet) has been reported to be anywhere from moderate-to-easy to propagate as softwood cuttings, according to Galle (1987). Treatment of Piedmont azalea softwood cuttings with 10,000 ppm K-IBA resulted in successful rooting performance (Knight et al., 2005). Also, rooting with lower rates of K-IBA (7500 ppm) occurred and an increase of rooting percentage was observed. Berry (1998) also observed rooting when new soft growth was treated with 5000 ppm K-IBA.

Transport of auxins occurs basipetally, from the shoot apices downward to the root apices (Goldsmith, 1977; Petrášek and Friml, 2009; Robert and Friml, 2009). Based on the location of naturally occurring auxins and the transport path, it could be possible to take new, young plant tissue cuttings and utilize these naturally occurring auxins. Addition of a low dose, synthetic auxin treatment to young deciduous azalea cuttings might encourage the plant to root faster and more efficiently than with a hardwood cutting. Successfully rooting softwood cuttings with lower doses of auxin could also potentially provide a solution to the dormancy problem. Thus, the objective of this study was to determine if rooting response was improved when softwood, single-node Piedmont azalea cuttings were treated with a basal quick dip in a range of IBA concentrations.

MATERIALS AND METHODS

A randomized complete block experimental design was utilized with eight cuttings per treatment, n=8. Piedmont azalea cuttings were taken on 3 April 2020 from a native population at The Crosby Arboretum located in Picayune, MS (lat. 30°30’11” N, long. 89°39’53” W, elevation 17 meters USDA hardiness zone 8b). Cuttings were taken around 7:00 am after a recent rain to ensure they were turgid to aid in reduction of transpiration stress on the cuttings (Bir, 1996). Using the method that was described by Jenkins (2007), the softwood cuttings were taken from tissue soft enough to be removed via pinching with fingers. Cuttings were single-node with an average length around 2.54 cm (1-in.). Immediately after pinching, cuttings were placed and stored in a cooler of water until being stuck with the respective treatments (Jenkins, 2007). At sticking, cuttings were turgid and showed no signs of wilting or stress.

Hortus (Hortus IBA Water Soluble Salts™, Phytotronics, Inc.®, Earth City, MO) was applied as a quick dip to the cuttings at four different auxin levels (0, 50, 250 or 400 ppm). Cuttings were then stuck in a growing mix (Jolly Gardner® Pro Line C/B Growing Mix, Old Castle Lawn & Garden, Atlanta, GA) in 5.7 cm (2.3 in.), 38 cell propagation tray inserts. The growing mix contained Canadian sphagnum peat, processed pine bark, coarse perlite, and medium vermiculite. Cuttings were then placed under intermittent mist for 6-sec every 6-min during daylight hours. Sixty days after sticking, mist intervals were reduced to 2-sec every
6- min. Data collected after 120 days included rooting percentage, total root number, root length (cm), root surface area (cm^2), average root diameter (mm), root volume (cm^3), number of root tips, number of root forks, number of root crossings, and root quality (1-5, with 1=no roots and 5=healthy, vigorous root system). Roots were washed by gently spraying with water, then were separated from the stem of each plant. The cleaned individual root systems were floated in tap water in a 10 by 15 cm (4- by 6-in.) Plexiglas tray and gray-scale root images were acquired. Roots were untangled and separated with a paintbrush to minimize root overlap. The tray was placed on top of a specialized dual-scan optical scanner, linked to a computer. Gray-scale root images were acquired by setting the parameters to high accuracy (resolution 800 by 800 dpi). Acquired images were analyzed for root length, root surface area, average root diameter, root volume, and number of tips, forks, and crossings using winRHIZO Pro software (Regent Instruments, Inc., Quebec, QC, Canada). Data were analyzed using PROC GLM and Tukey's HSD at $P \leq 0.05$ in SAS version 9.4 (SAS Institute INC., Cary, NC).

RESULTS AND DISCUSSION

Rooting percentage, total root number, and root quality were similar among treatments (Table 1). Rooting percentages ranged from 72 to 91% depending on treatment with overall rooting percentages of 83% across all treatments. In other studies, Piedmont azalea rooting percentages have ranged from 75 to 100% (Knight et al., 2005; Thompson, 2018). Even though rooting percentages are very similar, previous studies used older cuttings compared to this study and required high concentrations of IBA (7500 to 10000 ppm) to achieve these percentages.

Propagation methods that will provide the grower with a quality liner at maximum efficiency are highly beneficial for success of Piedmont azalea in the market. Based on the results in this study, it would appear that new growth, softwood, single node cuttings of Piedmont azalea can be successfully rooted without an additional exogenous auxin application. Successfully propagating hard to root species without the use of exogenous auxins can benefit propagation nurseries financially and in management practices. Plant production without the use of auxin can provide a savings in input costs spent on auxins and labor costs. Knowing that Piedmont azalea can be propagated without the use of auxin will reduce the time it takes to stick cuttings by eliminating the hormone dipping process. Growers can also reduce the spread of disease by being able to stick cuttings directly into growing media without the possibility of contaminating them in the hormone dipping process. Also, knowing how to propagate Piedmont azalea from different cutting types gives growers the flexibility to propagate at the most efficient time during their crop rotations.

Table 1. Influence of auxin basal quick dip on root percentage, number of roots, and root quality of Piedmont azalea.

Treatment	Rooting[w] (%)	Roots (no.)	Root quality rating[x]
Untreated	91a[z]	5.8a	3.6a
IBA			
50 ppm	88a	4.7a	3.3a
250 ppm	81a	4.8a	3.1a
400 ppm	72a	4.1a	2.7a
Significance (p-value)[y]	0.2318	0.3180	0.2411

[z]Means followed by the same letter within a column are similar and not significantly different ($\alpha = 0.05$).

[y]P values for differences between means were obtained using Tukey's honest significant difference (HSD) at $P \leq 0.05$.

[x]Root quality (visual rating of 0-5, with 0=dead, no callus and 5=healthy, vigorous root system).

LITERATURE CITED

Berry, J. (1998). Commercial propagation of southern native woody ornamentals. Comb. Proc. Intl. Plant Prop. Soc. *48*:643–650.

Bir, R.E. (1992). Growing and propagating showy native woody plants. Chapel Hill (NC): University of North Carolina Press.

Bir, R.E. (1996). Rooting stem cuttings of some eastern native *Rhododendrons*. J. Amer. Rhododendron Soc. 50 (2). https://scholar.lib.vt.edu/ejournals/JARS/v50n2/v50n2-bir.htm

Brown, B. (2017). Fungal associations and improving micropropagation of native *Rhododendron* spp. Master's Thesis, Auburn University.

Dirr, M.A. and Heuser, Jr.. C.W. (2018). The reference manual of woody plant propagation. Timber Press. Portland, Oregon.

Galle, F.C. (1987). Azaleas. Timber Press. Portland, Oregon.

Goldsmith, M. H. M. (1977). Polar transport of auxin. Annu. Rev. Plant Physiol. Plant Mol. Biol. *28*:439-478.

Hyatt, D.W. (2006). Propagation of deciduous azaleas. Comb. Proc. Intl. Plant Prop. Soc. *56*:542-547.

Jenkins, M.Y. (2007). Rooting stewartia and native azaleas using softwood cuttings. Comb. Proc. Intl. Plant Prop. Soc. *57*:646-647.

Knight, P., Coker, C., Anderson, J., Murchison, D. and Watson, C. (2005). Mist interval and K-IBA concentration influence rooting of orange and mountain azalea. Native Plants Journal. *6*:111-117.

Petrášek, J. and Friml, J. (2009). Auxin transport routes in plant development. Development *136*:2675-2688.

Robert, H.S. and Friml, J. (2009). Auxin and other signals on the move in plants. Nature Chemical Biology *5*:325-332.

Ryals, J. B., Knight, P.R., Chastain, D.R., Ryals III, L.E., Coker, C.E., Bachman, G.R., DelPrince, J.M., Drackett, P.R., and Bowden, A.T. (2019). Effect of cutting submersion duration and auxin concentration on survivability and root response of Florida azalea. Comb. Proc. Intl. Plant Prop. Soc. *69*:295-300.

Sommerville, E.A. (1998). Propagating Native Azaleas. J. Amer. Rhododendron Soc. *52*:126-127.

Thompson, P. (2018). Rooting response of deciduous azaleas, *Rhododendron* section *Pentanthera,* stem cuttings to mist regimes and media mixes. Master's Thesis, Auburn University.

Using Poinsettia and Pepper as Model Plants to Investigate Biochar and *Trichoderma* Suppressing Effects on Plant Diseases

Ping Yu[1], Kevin Ong[2], Kevin Crosby[1], Terry Gentry[3], and Mengmeng Gu[4]

[1] Department of Horticultural Sciences, Texas A&M University, College Station, TX, 77843, USA; [2] Department of Plant Pathology and Microbiology, Texas A&M University, College Station, TX, 77843, USA; [3] Department of Soil and Crop Sciences, Texas A&M University, College Station, TX, 77843, USA; [4] Department of Horticultural Sciences, Texas A&M AgriLife Extension Services, College Station, TX, 77843, USA

yuping520@tamu.edu

Keywords: Disease control, *Phytophthora capsica, Pythium aphanidermatum*, hardwood biochar

Summary

Biochar (BC) is a carbon-rich by-product from biomass pyrolysis (thermochemical biomass decomposition under an oxygen-depleted or oxygen-limited environment with specific time and temperature conditions). Biochar is of commercial importance for replacing more costly peat moss-based substrate for greenhouse plant production - and its potential to suppress plant diseases such as *Phytophthora capsici* and *Pythium aphanidermatum.* The application of *Trichoderma* did not significantly reduce disease severity. However, the mixed hardwood biochar (HB) mixed at 20% by volume could replace peat moss-based substrate to reduce poinsettia root rot disease caused by *P. aphanidermatum* without negatively affecting poinsettia plant growth. Incorporating HB by replacing 30% and 50% peat moss in the substrate could also reduce pepper blight disease caused by *P. capsici*.

INTRODUCTION

Potted poinsettia (*Euphorbia pulcherrima*) is one of the most important greenhouse ornamental crops in the United States, with an estimated wholesale value of $170 million in the top 15 states (USDA-NASS, 2018). Pepper is another important crop with a market value around $784 million (USDA-NASS, 2018). *Pythium aphanidermatum* and *Phytophthora capsici* are two common pathogens in greenhouses affecting poinsettia and pepper production significantly. They can both survive and thrive with high humidity and even high temperature. *P. aphanidermatum* is the predominant *Pythium* species causing poinsettia root rot disease, which is a recurrent disease that affects poinsettia production in greenhouses across the US (Lookabaugh et al., 2020; Múnera et al., 2019). *P. capsici* is a destructive hemi-biotrophic pathogen causing disease on a broad range of crops from families including solanaceous, cucurbitaceous, and fabaceous (Kousik et al., 2015). *Phytophthora* blight on pepper caused by *P. capsici* is one of the most serious soil-borne diseases for pepper worldwide (Wang et al., 2019).

Biochar (BC) is a carbon-rich by-product from biomass pyrolysis (thermochemical biomass decomposition under an oxygen-depleted or oxygen-limited environment with specific time and temperature conditions) (Demirbas and Arin, 2002; Lehmann, 2007). Not only can biochar replace peat moss-based substrate for greenhouse plant production (Guo et al., 2018; Huang et al., 2019; Yan et al., 2020; Yu et al., 2020) but it has the potential to suppress plant diseases including the diseases caused by *P. capsici* and *P. aphanidermatum.* For instance, BC-amended soil suppressed disease caused by *Pythium spp.* was reported, although with BC at relatively low rates (≤ 3% w/w) (Jaiswal et al., 2019). Also, incorporating corn stalk biochar (pH 9.73) at 13.7g/kg to soil suppressed pepper blight because it increased the abundance of beneficial microorganisms (Wang et al., 2020).

Trichoderma spp. has been reported as a reliable biological control agent for *P. capsici* and *P. aphanidermatum.* For instance, *Trichoderma harzianum* was proven to suppress pepper root rot caused by *P. capsici* through antimicrobial substances production (Ezziyyani et al., 2007). In a vitro test, *T. harzianum* inhibited *P. capsici* by 65.3% (Das et al., 2019). Also, *Trichoderma spp.* played a role in controlling cucumber damping-off caused by *P. aphanidermatum* (AL-Malikya et al., 2018). Some types of BC have proven to have synergistic effects with other components including *Trichoderma spp.* As such, we conducted two greenhouse trials using poinsettia and pepper plants as model plants to test the BC suppression effects on plant diseases caused by *P. capsici* and *P. aphanidermatum.*

MATERIALS & METHODS

Plants Material, Biochar Media, Pathogen & Trichoderma

Poinsettia (*Euphorbia pulcherrima* 'Prestige Sunrise Red') cuttings were stuck in commercial propagation media (BM2 Berger, Saint-Modeste, Quebec, Canada). After the root grew out, uniform cuttings were transplanted into 6-inch azalea pots filled with peat moss-based commercial substrate (CS, Jolly Gardener C/20, Oldcastle Lawn & Garden Inc., Atlanta, GA, USA) incorporated with mix hardwood biochar (HB, Proton Power Inc. Lenoir City, TN, USA) mixed at 20% and 40% (by vol.). Hot cherry pepper (*Capsicum annuum* 'Capperino') F1 plants were grown in the greenhouse and self-pollinated to get F2 seeds. According to Johnny's seeds, the F1 seeds are susceptible to *P. capsici* (personal conversation). Based on our two previous preliminary studies, there were no patterns of *P. capsici* resistance. Because F1 seeds were not *P. capsici* resistant, F2 plants showed no patterns on *P. capsici* resistance, and the difficulties of passing on the disease resistance to the descendants, we can safely assume that the F2 seeds used in this study are not *P. capsici* resistant. Uniform seedlings were transplanted into 4-inch pots filled with CS blended with either HB at 10%, 30%, 50%, and 70% (by vol.) or sugarcane bagasse biochar (SBB, American Biocarbon LLC White Castle, LA, USA) at 10% (by vol.). The CS used in this study contains 80% Canadian Sphagnum peat moss with the rest being perlite and was used as the control. At transplanting, slow release fertilizer Osmocote Plus (15N-4P-10K, Scotts-Sierra Horticultural Products Company, Marysville, OH, USA) was applied at manufactory rates.

Pythium aphanidermatum and *Phytophthora capsici* were isolated from infected plants and inoculated with actively growing mycelium agar. Root shield Plus-WP (BioWorks, Victor, NY, USA) was used as a biological control agent in this study. The biological control agent contained two active strains of *Trichoderma*, *T. harzianum* strain T-22, and *T. virens* strain G-41. *Trichoderma*-containing product was applied at the manufactory's recommenda-tion rate four weeks after plant transplanting (WK4, poinsettia, WK1 for pepper). This experiment was designed as random complete block design and was conducted in the greenhouses located on Texas A&M University campus, College Station (poinsettia) and Sommerville (pepper), Texas, USA. The average greenhouse temperature, relative humidity, and dew point were 30.2 °C, 77.2%, and 25.0 °C, respectively.

Measurements:

Potting mix physical and chemical properties:

Media physical properties—total porosity (TP), container capacity (CC), air space (AS), and bulk density (BD)—were measured according to North Carolina State University Horticultural Substrates Laboratory Porometer (Fonteno et al., 1995). The leachate electrical conductivity (EC) and pH were measured with a portable EC/pH meter (Hanna Instrument, Woonsocket, RI, USA), according to the pour-through method (LeBude and Bilderback, 2009).

Disease assessment:

Disease symptoms were observed and recorded every 5 days after pathogen inoculation. The disease severity was recorded at a 0-4 scale (no symptom - dead plants) according to Wang's work (Wang et al., 2019). The scale was also visualized in this work as shown in Fig. 1 and 2.

Disease severity index (DS) (Wang et al., 2019) was calculated by the following formula:

$$\text{DS} = \sum\left(\frac{\text{number of diseased plants in this index} \times \text{disease index rating from 0 to 4}}{4 \times \text{number of plants investigated}}\right) \times 100\%.$$

Figure 1. 0-4 scales (0 = no symptom, 4 = dead plant) used for the poinsettia root rot caused by *P. aphanidermatum* disease severity rating used in this study, no plant was dead in this study.

Figure 2. Visual scales (0-4; 0 = no symptom; 4 = dead) used for the pepper blight caused by *Phytophthora capcisi* disease severity rating used in this study.

Disease incidence (DI) was evaluated by counting the number of diseased plants in each pot twice during the trials, according to the formula:

$$\text{DI} = \frac{number\ of\ diseased\ plants}{number\ of\ total\ plants} \times 100$$

(Bellini et al., 2020).The disease severity obtained at different times after inoculation was used to calculate areas under disease progress curves (AUDPC) following the formula:

$$\text{AUDPC} = \sum_{i=0}^{n-1} \frac{(y_i + y_{i+1})}{2} (t_{i+1} - t_i).$$

Where y_i is the scale rating at the ith observation, $t_i$is the day of the ith observation, and n is the total number of observations (Madden et al., 2007).

RESULTS

Potting mix physio-chemical properties:

Most of the mixes' physical properties were within the recommended range (Yeager et al., 2007), except for the BDs in all the treatments, which were lower than the recommended value (Table1).

The HB20 and HB40 mixes had a significantly lower TP, and pH, as compared with the control (CS100). The HB50 and HB70 mixes had significantly lower TP, CC and BD, while they had significantly higher pH as compared with the control (CS100).

Table 1. Container compacity (CC), air space (AS), total porosity (TP), and bulk density (BD) of the commercial substrate (CS), mixed hardwood biochar (HB), and sugarcane bagasse biochar mixes (SBB).

Trt.	TP (%)	CC (%)	AS (%)	BD (g cm^{-3})	pH	EC (μS cm^{-1})
			Poinsettia			
CS100	74±0.3	56±0.2	18±0.5	0.09±0.00	6.8±0.05	2,058±29
HB20	72±0.3*	54±1.2	17±1.5	0.09±0.00	7.6±0.05***	2,022±26
HB40	70±0.5*	52±1.0	18±0.6	0.11±0.00**	8.2±0.01***	1,457±11***
			Pepper			
SBB10	73±0.1	61±1.7	13±1.6	0.10±0.00	6.6±0.03	1,065±72***
HB10	72±0.3	54±1.2	17±1.5	0.09±0.00	7.5±0.04***	1,960±18
HB30	70±0.5	52±1.0	18±0.6	0.11±0.00**	7.9±0.03***	1,830±32
HB50	68±3.0*	50±1.2*	18±4.0	0.12±0.00***	8.0±0.08***	1,575±178**
HB70	68±0.8*	47±1.5***	21±2.0	0.13±0.00***	8.4±0.10***	1,395±67***
Suitable range^z	50-80	45-65	10-30	0.19-0.7	5.4-6.5	<1,500

Note: Numbers after CS, SBB, and HB indicated the ratio of different components, by vol. *, **, and *** indicates significant difference from the commercial substrate (CS100) according to Dunnett's test at $p \leq$ 0.1,0.05, and 0.01, respectively. ^{z}Recommended physical properties of container substrate by (Nelson, 2012); Yeager et al. (2007).

Disease parameters

Disease symptoms of *Pythium* poinsettia root rot appeared in transplants in all the treatments at 5 days after inoculation (Fig. 3. A). Compared with CS100 treatments, HB20 treatments maintained a low disease severity throughout the experiment and reduced the disease severity at 5, 10, 15, 20, and 25 days after inoculation by 10.9%, 10.9%, 18.8%, 21.9%, respectively.

The application of *Trichoderma* did not significantly reduce disease severity throughout the experiment (Fig. 3. B). Also, HB20 treatments reduced disease incidence by 31.3% starting at 5 days after inoculation (data not shown).

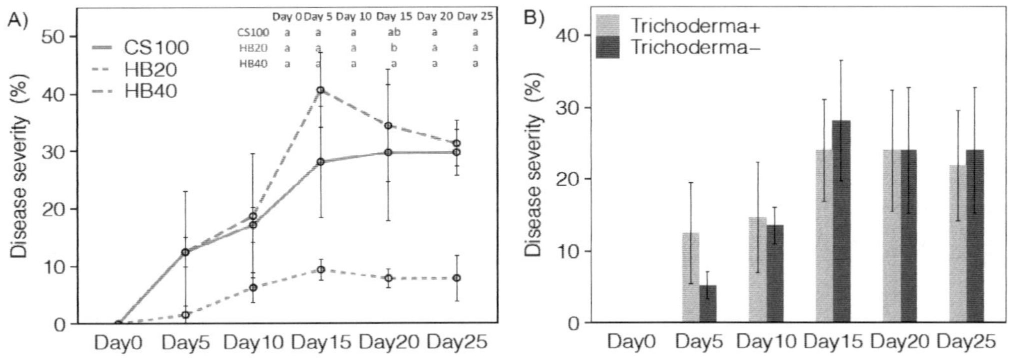

Figure 3. The effect of biochar rates (A) and *Trichoderma* (B) on disease severity for pathogen inoculate poinsettia plants. CS100 = peat moss-based commercial substrate, HB20 and HB40 = 20% and 40% (by vol.) mixed hardwood biochar-amended mixes, respectively. The same letter indicates not significantly different from each other according to LSD multiple comparison test at $p \leq 0.05$ on the same day.

Biochar mixes had significant impacts on pepper plants disease severity, especially HB-amended (30%-70%) mixes (Fig. 4. A). Compared with CS100 treatment, HB50 and HB70 treatments reduced disease severity at 12 days after transplanting by 10.94% and 10.16%, respectively. The HB 50 and HB 70 also significantly reduced disease incidence by 25.0%, respectively (data not shown).

The application of *Trichoderma* did not significantly reduced disease incidence the entire experiment (Fig. 4. B). All the BC-amended mixes had significantly lower AUDPC values (except for HB10) than the CS100 (data not shown).

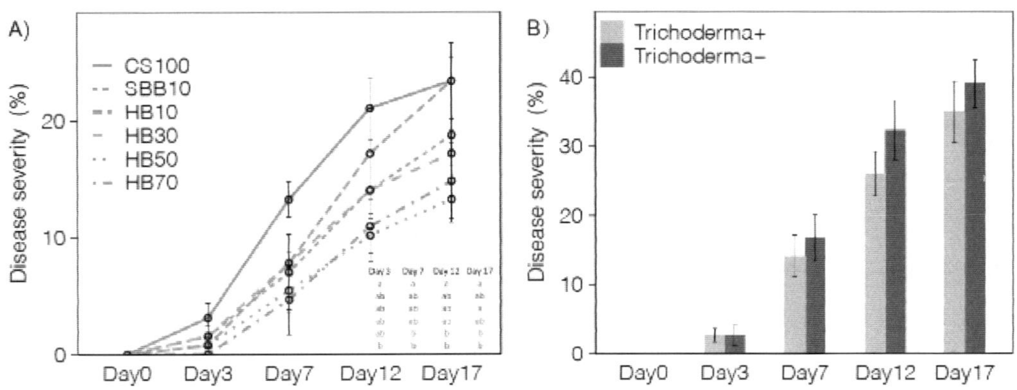

Figure 4. The effect of biochar rates (A) and *Trichoderma* (B) on disease severity for pathogen-inoculate treatments. SBB = Sugarcane bagasse biochar, HB = Mixed hardwood biochar, CS = Peat moss based commercial substrate. Numbers after CS, SBB, and HB indicated the ratio of different components, by vol. The same letter indicates not significantly different from each other on the same day according to LSD multiple comparison test at $p \leq 0.05$.

LITERATURE CITED

AL-Malikya, B.S., Jamela, D.S., Kareema, T. and AL-Sadaawyb, A.K. (2018). Effect of spent mushroom substrate (SMS) on cucumber damping off and root rot caused by *Pythium aphanidermatum* in greenhouse. Asian J. Microbio. Biotech. and Environ. Sci. *20*:S1-S5.

Bellini, A., Ferrocino, I., Cucu, M.A., Pugliese, M., Garibaldi, A. and Gullino, M.L. (2020). A compost treatment acts as a suppressive agent in *Phytophthora capsici–Cucurbita pepo* pathosystem by modifying the rhizosphere microbiota. Frontiers Plant Sci. *11*:885.

Das, M.M., Haridas, M. and Sabu, A. (2019). Biological control of black pepper and ginger pathogens, *Fusarium oxysporum*, *Rhizoctonia solani* and *Phytophthora capsici*, using *Trichoderma spp*. Biocatalysis and Agri. Biotech. 17:177-183.

Demirbas, A. and Arin, G. (2002). An overview of biomass pyrolysis. Energy Sources *24*:471-482.

Ezziyyani, M., Requena, M., Egea-Gilabert, C. and Candela, M. (2007). Biological control of Phytophthora root rot of pepper using *Trichoderma harzianum* and *Streptomyces rochei* in combination. J. Phytopath. *155*:342-349.

Fonteno, W., Hardin, C. and Brewster, J. (1995). Procedures for determining physical properties of horticultural substrates using the NCSU Porometer. Horticultural Substrates Laboratory, North Carolina State University.

Guo, Y., Niu, G., Starman, T. and Gu, M. (2018). Growth and development of Easter lily in response to container substrate with biochar. J. Hort. Sci. and Biotech. *94*:80-86.

Huang, L., Yu, P. and Gu, M.. (2019). Evaluation of biochar and compost mixes as substitutes to a commercial propagation mix. Applied Sciences *9*:4394.

Jaiswal, A.K., Graber, E.R., Elad, Y. and Frenkel, O. (2019). Biochar as a management tool for soilborne diseases affecting early stage nursery seedling production. Crop Protection *120*:34-42.

Kousik, C.S., Parada, C. and Quesada-Ocampo, L. (2015). First report of Phytophthora fruit rot on bitter gourd (*Mormodica charantia*) and sponge gourd (*Luffa cylindrica*) caused by *Phytophthora capsici*. Plant Health Progress *16*:93-94.

LeBude, A. and Bilderback, T. (2009). The pour-through extraction method: A nutrient management tool for nursery crops. NC State Univ. Coop. Ext. Bul. AG–717–W, North Carolina State University: Raleigh, NC, USA.

Lehmann, J. (2007). A handful of carbon. Nature *447*:143-144.

Lookabaugh, E.C., Kerns, J.P. and Shew, B. (2020). Evaluating fungicide selections to manage *Pythium* root rot on poinsettia cultivars with varying levels of partial resistance. Plant Disease *105*:1640-1647.

Madden, L.V., Hughes, G. and Van Den Bosch, F. (2007). The study of plant disease epidemics. American Phytopathological Society St. Paul.

Múnera, J.D.C., Belayneh, B., Ritsvey, A., Koivunen, E.E., Lea-Cox, J. and Swett, C.L.. (2019). Enabling adaptation to water scarcity: Identifying and managing root disease risks associated with reducing irrigation inputs in greenhouse crop production–A case study in poinsettia. Agricultural Water Management *226*:105737.

Nelson, P. (2012). Root substrate. Greenhouse operation and management. 7th ed. Prentice Hall, Upper Saddle River, NJ:161-194.

USDA-NASS. (2018). Agricultural Statistics, p. 202-210. In: U.S.D.o. Agriculture (ed.). United States Government Printing Office Seattle, WA, USA.

Wang, G., Govinden, R., Chenia, H.Y., Ma, Y., Guo, D. and Ren, G. (2019). Suppression of Phytophthora blight of pepper by biochar amendment is associated with improved soil bacterial properties. Biology and Fertility of Soils *55*:813-824.

Wang, G., Ma, Y., Chenia, H.Y., Govinden, R., Luo, J. and Ren, G. (2020). Biochar-mediated control of phytophthora blight of pepper is closely related to the improvement of the rhizosphere fungal community. Frontiers Microbio. *11*:1427.

Yan, J., Yu, P., Liu, C., Li, Q. and Gu, M. (2020). Replacing peat moss with mixed hardwood biochar as container substrates to produce five types of mint (*Mentha spp.*). Industrial Crops and Products *155*:112820.

Yeager, T., Fare, D., Lea-Cox, J., Ruter, J., Bilderback, T., Gilliam, C., Niemiera, A., Warren, S., Whitewell, T. and White, R. (2007). Best management practices: Guide for producing container-grown plants. Southern Nursery Association. Marietta, GA.

Yu, P., Li, Q., Huang, L., Qin, K., Niu, G and Gu, M. (2020). The effects of mixed hardwood biochar, mycorrhizae, and fertigation on container tomato and pepper plant growth. Sustainability *12*:7072.

Identification of New Crapemyrtle Bark Scale (*Acanthococcus lagerstroemiae*) Hosts (*Spiraea* and *Callicarpa*) through DNA Barcoding

Runshi Xie[1], Bin Wu[1], Mengmeng Gu[2], Stacey R. Jones[3], James Robbins[4] and Hongmin Qin[5]

[1] Department of Horticultural Sciences, Texas A&M University, College Station TX 77843, USA; [2] Department of Horticultural Sciences, Texas A&M AgriLife Extension Service, College Station, TX 77843, USA; [3] North Carolina Cooperative Extension, North Carolina State University, Concord, NC 28027, USA; [4] Division of Agriculture, University of Arkansas, Little Rock, AR 72204, USA; [5] Department of Biology, Texas A&M University, College Station 77840, USA

fushe001@tamu.edu

Keywords: Scale insect, CMBS, host range, phylogeny, species identification

Summary

Global trade and international travel have led to the establishment of invasive pests in territories outside the pests' normal range. We have been following the distribution of crapemyrtle bark scale (CMBS; *Acanthococcus lagerstroemiae*), an invasive insect first discovered in the United States in 2004. In addition to the rapid geographical expansion of the CMBS distribution, one crucial concern is its ability to infest a wide range of plant species, beyond its primary host - *Lagerstroemia*. By studying the molecular evidence, we revealed the genetic relationships of CMBS specimens from different geographical locations and hosts. Naturally occurring CMBS infestations were confirmed on a native plant species, American beautyberry (*Callicarpa americana* L.), and Spirea (*Spiraea* L.) in the United States. The new infestation of CMBS found on *Spiraea* raises the alarm for the green industry that other economically important crops in the Rosaceae family might potentially be susceptible to CMBS attacks.

INTRODUCTION

Crapemyrtle bark scale (CMBS; *Acanthococcus lagerstroemiae*) is exotic pest insect that has been raising concerns since its first discovery in the United States in 2004. To date, CMBS was found in at least 17 states, from south central to the east coast of the United States, with the higher number of infestation reports found in Texas (115), Oklahoma (34), Arkansas (30), Louisiana (29). Mississippi (25), North Corallina (23), and Virginia (23) (EDDMapS., 2021). Additional to the rapid geographical expansion of the insect distribution, one crucial issue with regards to CMBS control and management is its polyphagous feeding habit which allows it to infest a wide range of plant hosts.

According to previous literatures and online insect database (such as Scalenet), which has accumulated a significant amount of host information for CMBS, plant species from at least 22 genera (15 families) has been listed as CMBS hosts (García Morales M, 2016). As the distribution of CMBS continues to expand beyond its native regions, specifically in the United States, concerns have been raised regarding the expanded host range for CMBS beyond *Lagerstroemia*, and the potential threats that CMBS poses to the native and economic important plants in the United States.

For instance, in 2018, *Hypericum kalmianum* L. (St. Johnswort) was confirmed by Schultz et al as CMBS hosts (Schultz & Szalanski, 2019). We have conducted CMBS feeding preference studies, to confirm previous documented hosts and identify potential new hosts, especially native plants, in the United States. From 2016 to 2019, we have confirmed several new species and cultivars as CMBS alternative hosts, including apple (*Malus domestica*), *Chaenomeles speciosa*, *Disopyros rhombifolia*, *Heimia salicifolia*, *M. angustifolia*, twelve pomegranate cultivars (Xie et al., 2020) and nine *Callicarpa* species (Wu, 2021).

In 2020, a scale infestation (suspected to be a CMBS infestation) observed on *Spiraea japonica* (Figure 1 A-B) at University of Arkansas. Again, in 2020, another incident of unknown scale infestation, later identified as CMBS, was reported by a homeowner in Concord, North Coralina, USA. The infested plant was identified as *Spiraea thunbergia* (Figure 1 C-D), which was planted sometimes between 1953 and 2016, and for long time without scale infestation. The scale infestation was believed to be initiated after two crapemyrtle plants infested with CMBS were accidentally planted nearby the *S. thunbergia*.

Spiraea, or commonly known as spirea or meadowsweet, is a genus consisting of 100 to 120 species known worldwide, primarily found in the temperate region of the northern hemisphere (Wrońska-Pilarek, Wiatrowska, & Bocianowski, 2019). The phylogeny of Rosaceae has revealed a closer relationship between Spiraeeae and Maleae (both under subfamily Amygdaloideae), while further from Rubeae (subfamily Rosoideae) (Xiang et al., 2017). Interestingly, this phylogenetic relationship is consistent with our previous findings, as CMBS infestation was found on apple and flowering quince (Chaenomeles) but not on blackberry and raspberry (Rubus) (Xie et al., 2020).

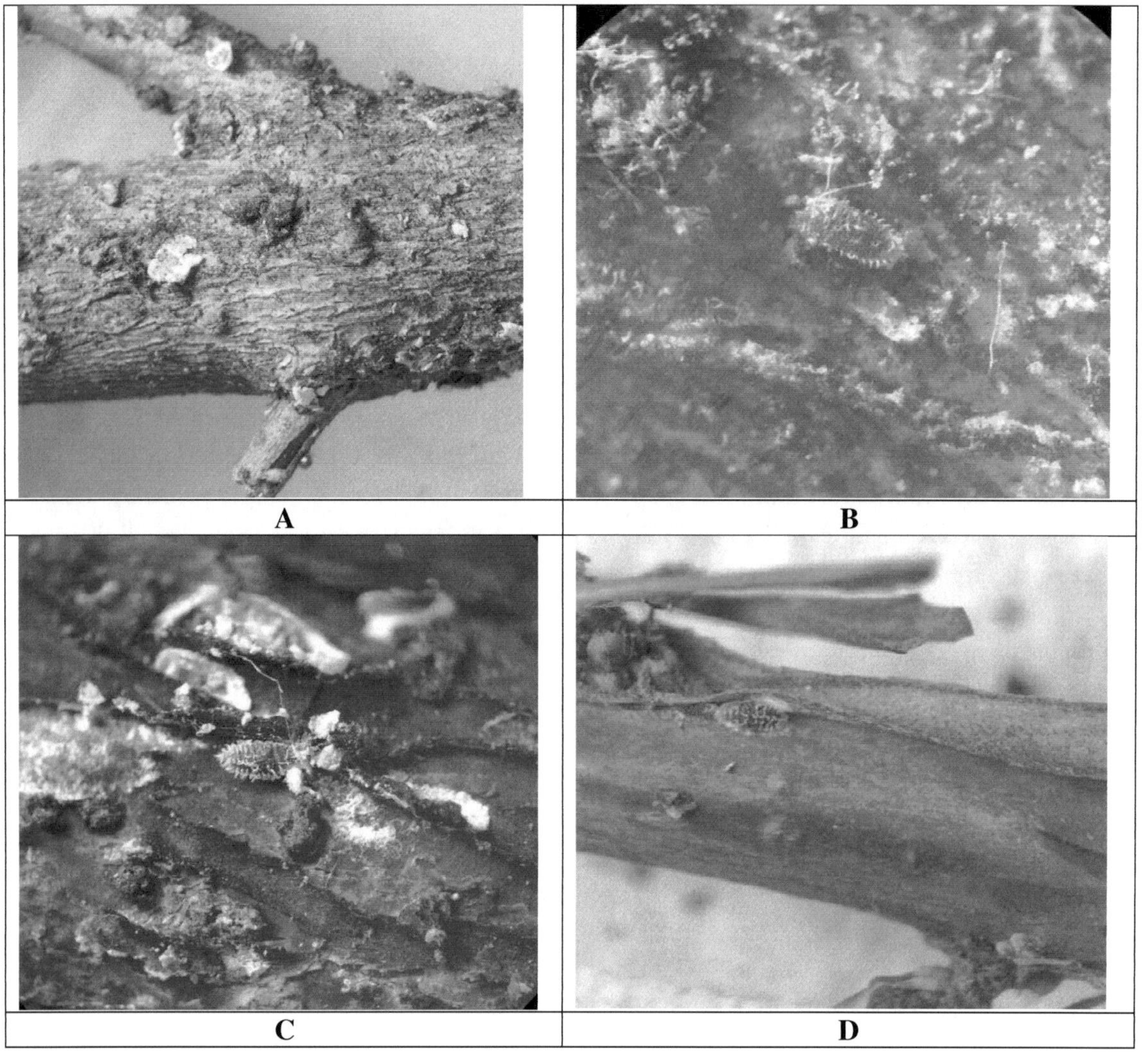

Figure 1. Microscopic images of crapemyrtle bark scale found on alternative hosts: (A-B) *Spiraea japonica*; (C-D) *S. thunbergia.*

Although the insect morphology suggested the unknown scale on *Spiraea* were CMBS, observational evidence still has limitations in identifying species. For example, the azalea bark scale, a close relative of CMBS, is often mistaken as CMBS since the morphology closely resembles each other. Therefore, to confirm the identity, molecular approaches were used to verify that CMBS indeed caused the infestation found on *S. japonica* and *S. thunbergii*.

MATERIALS AND METHODS

Insect sample collection and handling

Naturally occurred scale infestation (suspected to be caused by CMBS) was found on *C. americana* (from Faulkner County, Arkansas), *S. japonica* (from Faulkner County, Arkansas), and *S. thunbergii* (from Concord, North Carolina). Live nymphs were then collected, using a fine pin or a fine brush, from infested twigs or branches of crapemyrtle (control) and alternative hosts (*Callicarpa* and *Spiraea*). Insects collected were used immediately or stored in -80°C freezer. DNA extractions were performed using DNeasy Blood and Tissue Extraction Kit (Qiagen Inc., Valencia, CA).

Molecular identification

To amplify DNA sequence more specifically from the CMBS COI gene, primers based on the CMBS COI sequence (GenBank accession number: AB439520.1) were developed. Primer pair: Forward: 5'-CCAGGATTTGGATTAATATCAC-3', and Reverse: 5'-TGAACCAATTGATGATAGAG-3', was designed and successfully amplified CMBS COI sequences with lengths around 640 bp in this study.

The PCR program used is followed: initially holding sample at 98 °C for 30s, then 30 cycles of 98 °C for 10s, 51 °C for 30 s, and 72 °C for 30s, followed by a final extension of 72 °C for 10 min. The successful PCR amplifications were checked on a 1% agarose gel following electrophoresis and visualized using a Gel Doc E.Z. imaging system (BIO-RAD, Inc., Hercules, CA). The PCR products were purified and sent to a sequencing lab (Eton Bioscience, Inc., San Diego, CA) for direct sequencing in both directions.

Data analysis

A 475 bp section of COI sequence was used to analyze the relationship of CMBS from different hosts and two close related scale insects (*Acanthococcus azaleae*, and *A. abeliceae*). Crapemyrtle bark scale sequences used in this study have been deposited in GenBank (GenBank accession number: MZ312637:MZ312640). A phylogenetic tree was constructed using BLAST pairwise alignments (Johnson et al., 2008).

RESULTS AND DISCUSSION

Sequencing results reveal that amplicons from four insect samples (from crapemyrtle, *Callicarpa*, *Spiraea japonica*, and *S. thunbergia*) had sequence length above 600 bp (601~664). The BLAST results showed sequences from each sample had identity scores above 98% matched with CMBS COI sequences in the database (GenBank accession number AB439520.1).

To study the relationship of CMBS collected from different geographical locations and plant hosts, we constructed a molecular phylogenetic tree to reveal the genetic relationship of CMBS relative to its two close relatives from the United States or Asia. The phylogenetic tree showed a clear separation between CMBS and azalea bark scale while sharing a common evolutionary origin compared to *Acanthococcus abeliceae* Kuwana (Fig. 2). All the CMBS collected from either crapemyrtle or alternative hosts (*Spiraea* and *Callicarpa*) in the U.S. location were grouped, versus CMBS from the Asia location, suggesting CMBS at different geographical areas might have been evolving differently.

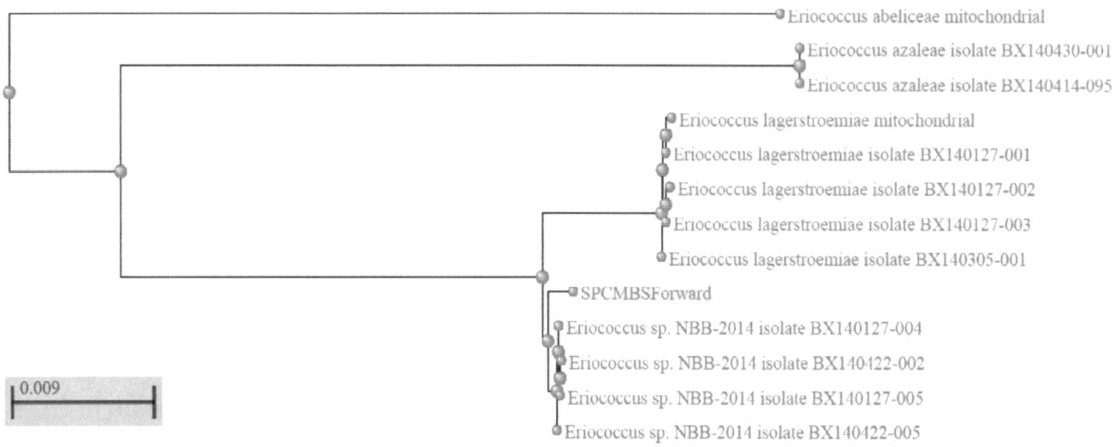

Figure 2. Phylogenetic tree shows the genetic relationship of *Acanthococcus* (=*Eriococcus*) *lagerstroemiae* (crapemyrtle bark scale from alternative hosts in this study were highlighted), *Acanthococcus azalea* (azalea bark scale), and *Acanthococcus abeliceae* from different geographical locations.

Our results have confirmed that, in addition to the geographical expansion, the expanding host range of CMBS is not only a potential threat to the green industry but a reality. The latest findings have brought the total number of CMBS hosts in Rosaceae to eight, making it one of the most prominent families hosting CMBS beside Lythraceae. According to the phylogeny within Rosaceae, confirmed CMBS host genera *Chaenomeles*, *Malus* and *Spiraea* had a closer genetic background with at least 55 genera under subfamily of Amygdaloideae (Potter et al., 2007), which includes many economically important crops such as apricot, almond, cherry, plum, and peach. Therefore, further investigation is needed for the potential threat of CMBS to other valuable crops in Rosaceae.

ACKNOWLEDGEMENTS

This work is supported by the Specialty Crop Research Initiative project 'Systematic Strategies to Manage Crapemyrtle Bark Scale, An Emerging Exotic Pest' [grant no. 2017-51181-26831/project accession no. 1013059] from the U.S. Department of Agriculture (USDA) National Institute of Food and Agriculture. Any opinions, findings, conclusions, or recommendations expressed in this publication are those of the author(s) and do not necessarily reflect the view of the U.S. Department of Agriculture.

LITERATURE CITED

EDDMapS. (2021). Early Detection & Distribution Mapping System - Center for Invasive Species and Ecosystem Health. Retrieved March 4, 2021, from The University of Georgia. http://www.eddmaps.org/

García Morales, M., D. B., Miller, D.R., Miller, G.L., Ben-Dov, Y., and Hardy, N.B. (2016). ScaleNet: A literature-based model of scale insect biology and systematics. Publication no. 10.1093/database/bav118. Retrieved January 30th, 2021 http://scalenet.info/catalogue/Acanthococcus%20lagerstroemiae/

Johnson, M., Zaretskaya, I., Raytselis, Y., Merezhuk, Y., McGinnis, S. and Madden, T. L. J. N. a. r. (2008). NCBI BLAST: a better web interface. *36*(suppl_2), W5-W9.

Schultz, P. B. and Szalanski, A. L. (2019). *Hypericum kalmianum* (St. Johnswort) Confirmed as a new host of the crapemyrtle bark scale in Virginia, USA. J. Agri. Urban Entomology *35*: 12-15.

Wrońska-Pilarek, D., Wiatrowska, B. and Bocianowski, J. (2019). Pollen morphology and variability of invasive *Spiraea tomentosa* L. (Rosaceae) from populations in Poland. PloS one *14*(8): e0218276.

Wu, B., Xie, R., Knox, G., Qin, H. and Gu M. (2021). Evaluating beautyberry and fig species as potential hosts of invasive crapemyrtle bark scale (Hemiptera: Eriococcidae). HortTechnology: 32: 10.21273/HORTTECH04897-21

Xiang, Y., Huang, C.-H., Hu, Y., Wen, J., Li, S., Yi, T., Chen, H. Wiang, J and Ma, H. (2017). Evolution of Rosaceae fruit types based on nuclear phylogeny in the context of geological times and genome duplication. Mol. Bio. Evolution *34*:262-281.

Xie, R., Wu, B., Dou, H., Liu, C., Knox, G. W., Qin, H. and Gu, M. (2020). Feeding preference of crapemyrtle bark scale (*Acanthococcus lagerstroemiae*) on different species. Insects *11*:399.

Camellia Propagation from Cuttings

Robert Black

Bennett's Creek Nursery, 18047 Sawyer Farm Lane, Smithfield, VA 23430 USA

Bob@bcnursery.com

Keywords: *Camellia japonica, Camellia sasanqua,* disbudding, herbicides, Hortus IBA Water Soluble Salts®, semi-hardwood cuttings, Jet-Ag®, propagation systems, production systems, pruning, vapor pressure deficit (VPD) controller, weed management

Summary

This paper describes the protocol for production of camellias from rooted cuttings. *Camellia japonica, C. sasanqua*, and *C.* hybrids are propagated by semi-hardwood cuttings. Cuttings are fully submerged in a 114 L (30-gal) tank of Jet-Ag® Solution (hydrogen peroxide, and peroxyacetic acid) to prevent diseases such as *Pythium* and *Phytophthora*. The semi-hardwood cuttings are trimmed to three or four leaves, approximately 15 cm (6-in.) in length, with green mottled tan or solid tan stems. The lower leaf is removed leaving a node about 2.5 cm (1-in.) from the base. No wounding is necessary. Bundles of cuttings are basal quick-dipped for 5-sec with 8,000 ppm IBA solution using Hortus IBA Water Soluble Salts® (20%). Misting is controlled with a Phytotronics® VPD clock to maintain well-hydrated unrooted cuttings. The misting system is composed of Tavlit® 866 mini-compact sprinklers. Mist applications are significantly reduced upon root initiation. Four weeks after root initiation, both *C. sasanqua* and C. japonica cultivars will have rooted, and misting is discontinued. *Camellia japonica* cultivars require six to eight weeks to initiate rooting.

INTRODUCTION

Bennett's Creek Nursery is a wholesale container growing operation located in Smithfield, Virginia (USDA hardiness zone 8A). A wide variety of plant material is produced on the 150 ha (400-acre) site (Fig. 1). Container production of woody ornamental shrubs, flower and shade trees, herbaceous perennials, and seasonal bedding plants are the main crops (Fig. 1).

Figure 1.
(Top) Aerial photograph of Bennett's Creek Nursery Smithfield location, and (bottom) camellia production in containers.

Among the ornamental shrubs are forty varieties of camellias: *C. japonica, C. sasanquas*, and a few hybrids are produced (Fig. 2). This paper describes the protocol for production of camellias from rooted cuttings.

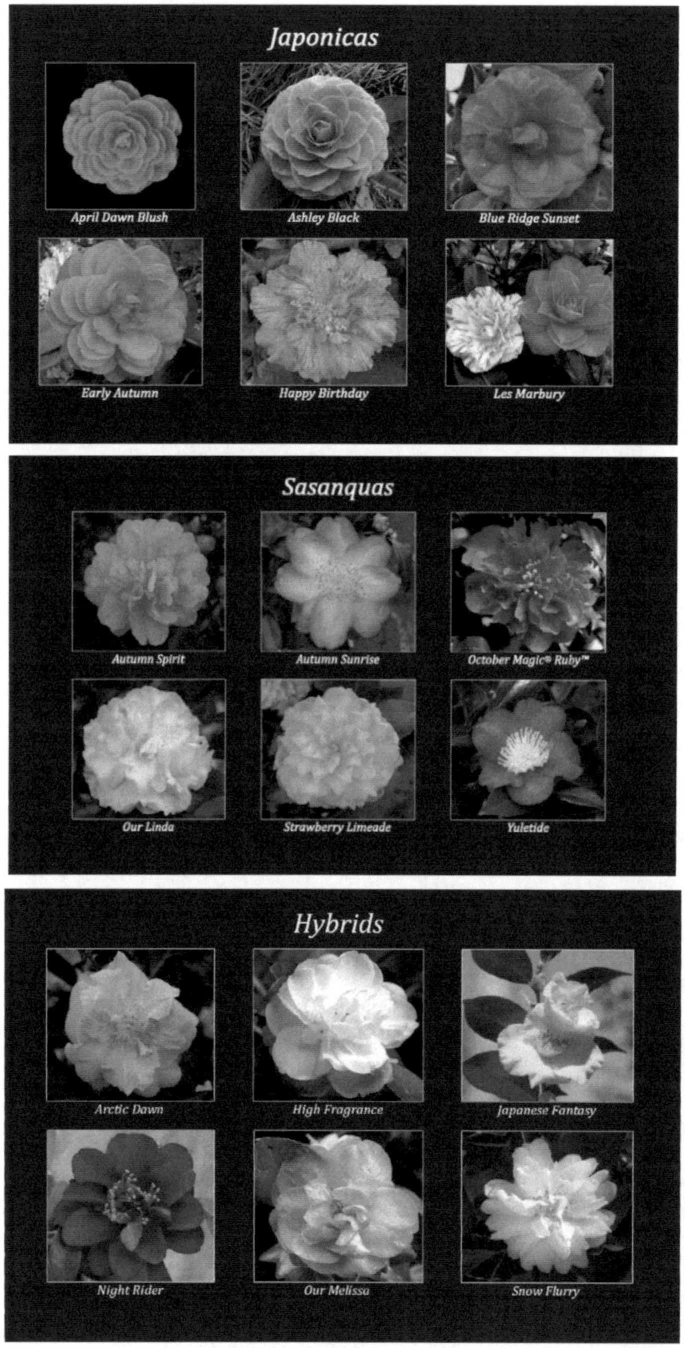

Figure 2. Images of the blooms of a portion of camellias grown at Bennett's Creek Nursery: [*C. japonica* (top), *C. sasanqua* (middle), and *C. hybrida* (bottom).

TIMING, CUTTING AND PREPARATION

All cuttings are collected on site from July through October. A production spreadsheet indicates the location and quantity of cuttings of each cultivar to cut and bundle (Fig. 3).

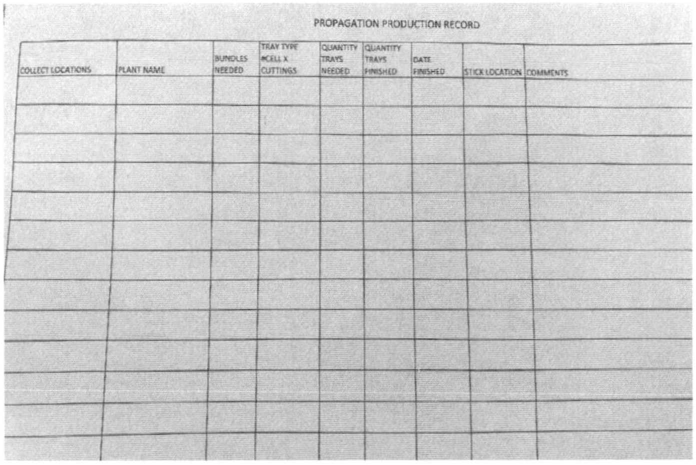
PROPAGATION PRODUCTION RECORD

COLLECT LOCATIONS	PLANT NAME	BUNDLES NEEDED	TRAY TYPE #CELL X CUTTINGS	QUANTITY TRAYS NEEDED	QUANTITY TRAYS FINISHED	DATE FINISHED	STICK LOCATION	COMMENTS

Figure 3. Propagation production spreadsheet at Bennett's Creek Nursery.

The cuttings are semi-hardwood with three or four leaves, approximately 15 cm (6-in.) in length, with green mottled tan or solid tan stems (Fig. 4). The lower leaf is removed leaving a node about 2.5 cm (1-in.) from the base. No wounding is necessary.

Cuttings are bundled into groups of 20 or 25 depending on the size of the cuttings (Fig. 4).

Figure 4. (left) Examples of *Camellia japonica* cuttings; (right) bundle of cuttings.

Bundles are crated and draped with damp burlap for protection from sun and desiccation (Fig 5).

The full crates are stored in a 2.7 m (12-ft) mobile air-conditioned trailer at the collection site (Fig. 5).

Figure 5. (left) Crated bundles protected with damp burlap, and (right) a propagation crew's van and air-conditioned trailer for storage and transport of cuttings.

Next, the crates are transported to the propagation facility for pre-stick dip treatment. Crates are fully submerged in a 114 L (30-gal) tank of Jet-Ag® Solution (a.i. Hydrogen Peroxide, and Peroxyacetic Acid Acid) to prevent diseases such as Pythium and Phytophthora (Fig. 6).

For Jet-Ag®, 30 ml (1-oz) is added to each 3.8 L (1-gal) gallon of water. The tank is periodically skimmed for debris removal and refreshed daily. Each week the tank is drained and replaced. Discarded Jet-Ag® solution is strained of debris then applied as a sanitizing agent to the floors of empty greenhouses.

Figure 6. (left) The oxidizing agent, Jet-Ag®, (right) a 19 L (5-gal) container of the hydrogen peroxide-based Jet-Ag® solution with a submerged crate of camellia bundles.

AUXIN TREATMENT

An 8,000 ppm IBA solution is prepared using Hortus IBA Water Soluble Salts® (20%) and distilled water. Bundles are basal quick-dipped for 5-sec, placed back into the crate, draped with damp burlap, and color coded with ribbon tied across the prepared crate (Fig. 7). The colored ribbons are a good visual aid for preventing mixing of cultivars (Fig 7).

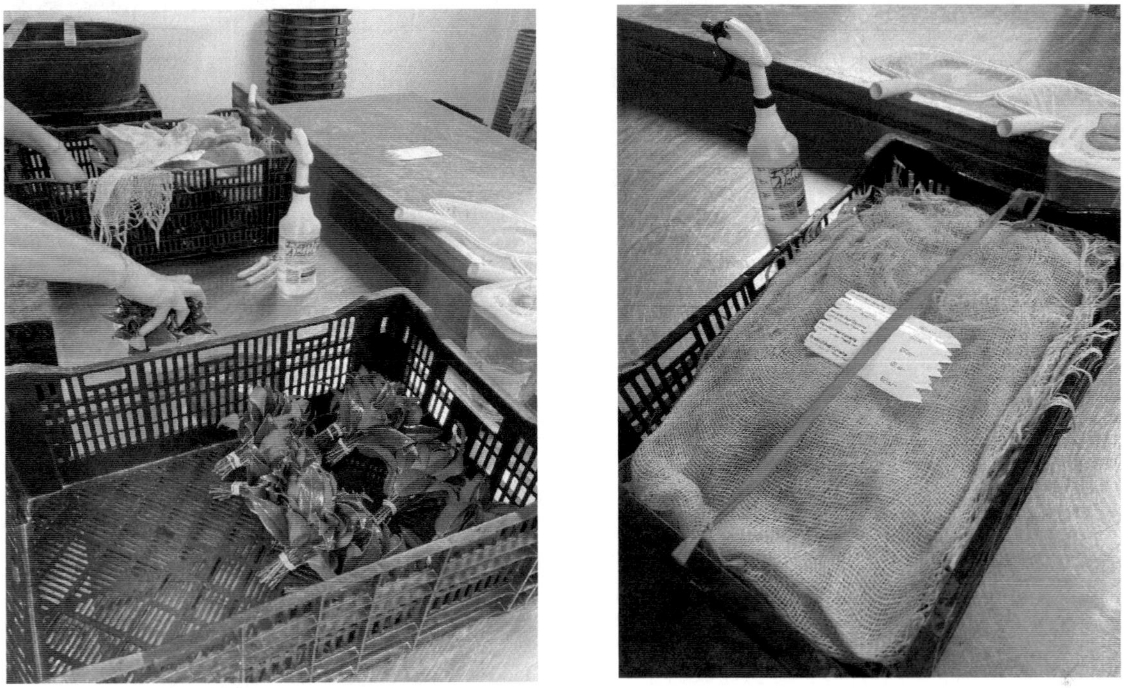

Figure 7. (left) Basal dipping bundles of cuttings with auxin solution, and (right) a finished crate of auxin-treated cuttings ready to stick.

SUBSTRATE

The components of the propagation rooting substate are detailed in Fig. 8. The rooting substrate is prepared daily in a 3.8 m^3 (5 yd^3) paddle mixer. A front loader is used to load the bulk aged pine bark (Fig. 9).

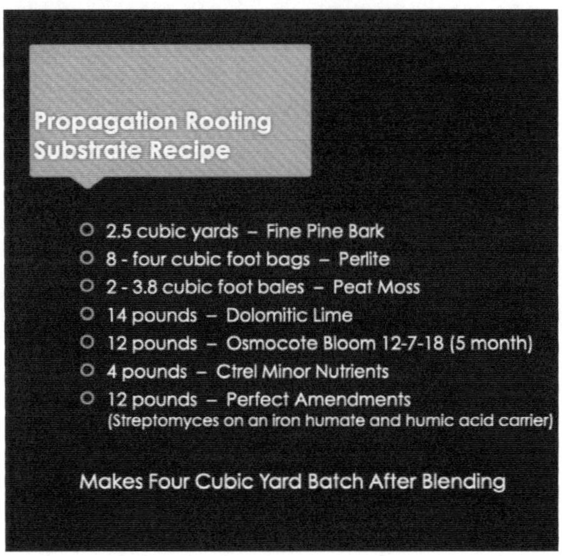

Figure 8. Blended ingredients to produce the rooting substrate.

Figure 9. (left) Loading course perlite into 3.8 m^3 (5 yd^3) paddle mixer, and (right) loading bulk aged loblolly pine bark into the paddle mixer.

FLAT FILLING

Twenty-one cell trays are filled with the rooting substrate inside the propagation building. As they exit the filling process, they are triple stacked on the trailer for transport to the greenhouses (Fig. 10).

Figure 10. Filling trays with rooting substrate.

STICKING TECHNIQUE

Cuttings are direct stuck, two per cell (Fig. 11). Emphasis is on proper depth and contact of the cutting stem with the media. Maximum depth is 4 cm (1.5-in.) with solid contact in the substrate. Each tray is tagged with the cultivar name.

Figure 11. (left) A propagation crew direct sticking into 21-cell trays, and (right) a tray of cuttings recently stuck.

ROOTING ENVIRONMENT

During the summer, greenhouses are covered with shade cloth and side curtains to provide adequate protection from direct sunlight and drying winds (Fig. 12). In late September, greenhouses are converted for cool season propagation (Fig. 13).

Ambient air temperature is maintained below 29 C° (85°F) by a thermostatically controlled vent fan. In addition, each greenhouse is equipped with a propane fired hot water in floor heating system to maintain 21 C° (70°F) in the rooting substrate (Fig 13).

Figure 12. (left) A 9x44 m (30x145 ft) greenhouse covered with 50% shade cloth and 2 m (6-ft) plastic side curtains. (right) Applying a double layer of clear plastic to greenhouse using boom attachment.

Figure 13. (left) Greenhouses converted for cool season propagation, and (right) wall mounted propane fired hot water heating system.

MISTING SYSTEM

Misting applications to maintain well hydrated unrooted cuttings is controlled by a Phytotronics® VPD clock (Fig. 14). Misting frequency is automatically adjusted based upon a continuous monitoring system. Ambient air temperature, relative humidity, and leaf surface temperature data is collected every ten seconds and are used to determine the vapor pressure deficit (VPD) (Fig. 15). The propagator enters a target based upon the level of moisture desired on the cuttings. Once VPDs are accumulated to the target, a misting event occurs. As temperature, humidity, and leaf surface temperature increases, VPDs are accumulated faster, and the misting frequency is increased.

The misting heads (Mini-Sprinklers) in the greenhouse are mounted inverted and overhead on weighted drop tubes (Fig. 15.). This design allows for an open floor area and accommodates unobstructed access for workers and equipment. The Mini-Sprinklers are manufactured by Tavlit®. They are the 866 mini-compact sprinklers and can accommodate a variety of color-coded spinners which have different flow-rates (Fig. 16). An anti-drip device is included in the design to quickly discontinue the misting event. It is engineered to open at 21 psi and close at 10 psi.

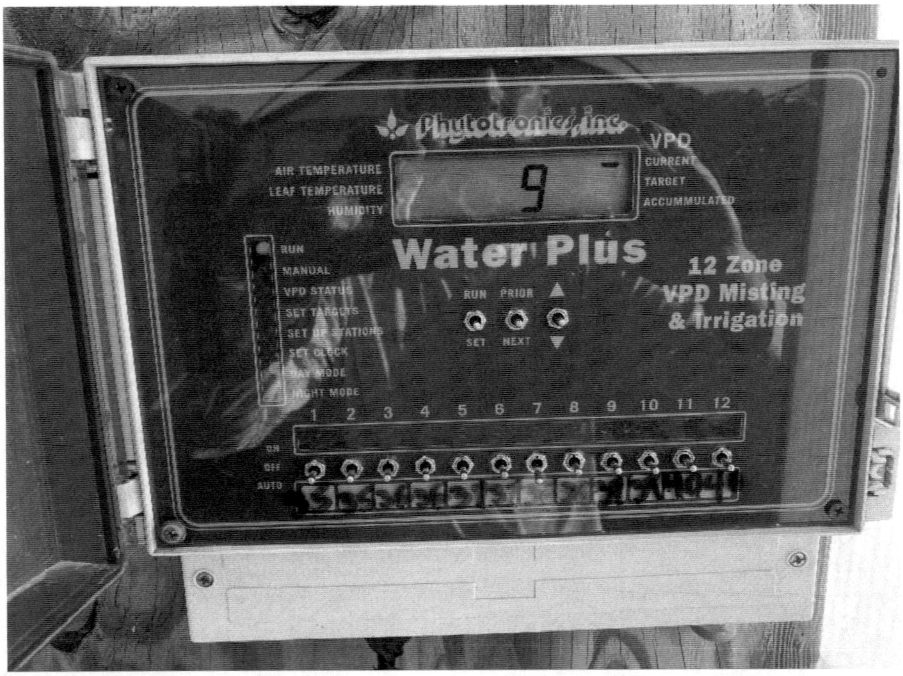

Figure 14. Phytotronics 12-zone vapor pressure deficit (VPD) controller.

Figure 15. (left) A data collector unit, and (right) a propagation house of recently stuck camellia cuttings with misting heads (Mini-Sprinklers) mounted overhead and inverted on weighted drop tubes (arrow).

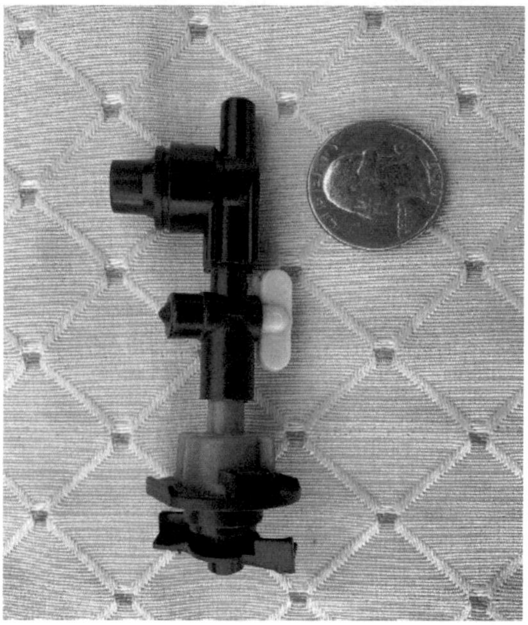

Figure 16. (Top, bottom left) Mini-Sprinklers are manufactured by Tavlit®. They are the 866 mini-compact sprinklers and can accommodate a variety of color-coded spinners which have different flow-rates. (Bottom right) An anti-drip device is included in the design to quickly discontinue the misting event.

WATER TREATMENT

Surface water is treated and supplied to the propagation greenhouses. It is filtered, acidified (target pH of 6.2), and chlorinated (target 2-ppm free chlorine). The irrigation system is computer controlled, variable flow, and maintains 55 psi. Acid and chlorine injection are also computer controlled to maintain the targets at various flow rates (Fig. 17).

Figure 17. (top left) Interior view of pump house, (top right) a Yardney sand filtration system with automatic backflush, (bottom left) chlorine gas cylinders with a regulator, and (bottom right) a Wallace and Tiernan® S10k gas feed chlorine metering system.

SPRAY PROGRAM

Unrooted cuttings are sprayed on a seven-day rotation to prevent algae, insect, and disease (Fig. 18). Maximum time between the spray application and mist applications resuming is accomplished by spraying after the final mist application for the day. Rotation between chemical classes (mode of action) prevents resistance development by insects and diseases. Once cuttings are rooted scouting determines application needs. In general, fully rooted cuttings are sprayed on a 30-day rotation to maximize growth rate and health.

Figure 18. Spray rotation tank mixes and applicator applying preventative spray to unrooted cuttings.

ROOTING TIMELINE

The time required for root initiation varies based upon camellia species and cultivars. *Camellia sasanqua* cultivars initiate rooting in three to four weeks on average. *Camellia japonica* cultivars require six to eight weeks to initiate rooting. Mist applications are significantly reduced upon root initiation. Four weeks after root initiations, both *C. sasanqua* and *C. japonica* are well rooted, and misting is discontinued (Fig 19).

Figure 19. A 9x44 m (30x145 ft) greenhouse of rooted cuttings.

FERTILITY PROGRAM

The fertilizer components incorporated into the rooting substrate provide a portion of the essential nutrients required for healthy root development. Upon root initiation, additional fertilizer is applied to supplement the incorporated nutrients (Fig. 20). Water soluble fertilizer formulations can be applied through the irrigation system or as a drench application. Alternating 20-10-20 and 17-5-17 NPK will provide all the essential nutrients. One application of each, fourteen days apart, at 200 ppm nitrogen will be sufficient going into the fall and the dormant season.

The following spring, fertilizer applications resume. A broadcasted application of a granular or controlled release fertilizer is an effective way to provide a steady low rate of nutrients to the young, rooted cuttings. A controlled release fertilizer, such as 15-9-12 NPK with a 5-to-6-month release pattern, at 1.4kg/9m^2 (3 lb/100 ft^2) works well.

A complete fertility analysis of both substrate and plant tissue can determine the nutritional status of the crop and help fine tune the fertilizer program. A water analysis is also helpful because some nutrients may be supplied by the water itself.

Following the Virginia Tech pour-through protocol, a portable electrical conductivity meter is used onsite to quickly monitor the pH and fertility level in the substrate (Fig. 21). A good target range for proper pH is 5.8 to 6.2 and electrical conductivity range for a good fertility indicator is 0.8 to 1.5 when growing camellia liners. To learn more about this, the pour-through method developed by Dr. Robert Wright review document AG-717WAWA.pdf at www.nurserycropscience.info.

Figure 20. Drench application of water-soluble fertilizer to rooted cuttings to encourage healthy root development.

Figure 21. (left) Applying water to rooted cutting in order to provide leachate for analysis, and (right) measuring electrical conductivity and pH data of leachate to determine fertility level.

OVERWINTERING

Once cuttings are well rooted and mist applications have been discontinued, the temperature in the greenhouse is gradually reduced over several weeks. This allows the plants to acclimate to dormant season temperatures without cold injury. Acclimated plants are able to withstand near freezing temperatures without injury. A minimum temperature at 2°C (35°F) in the greenhouse will provide adequate protection to well acclimated rooted cuttings (Fig. 22). Upon acclimation, the rooted cuttings are transferred to an overwintering house. White overwintering plastic provides 50% shade during the dormant season. The white plastic is removed and replaced with 50% shade cloth after the frost-free date.

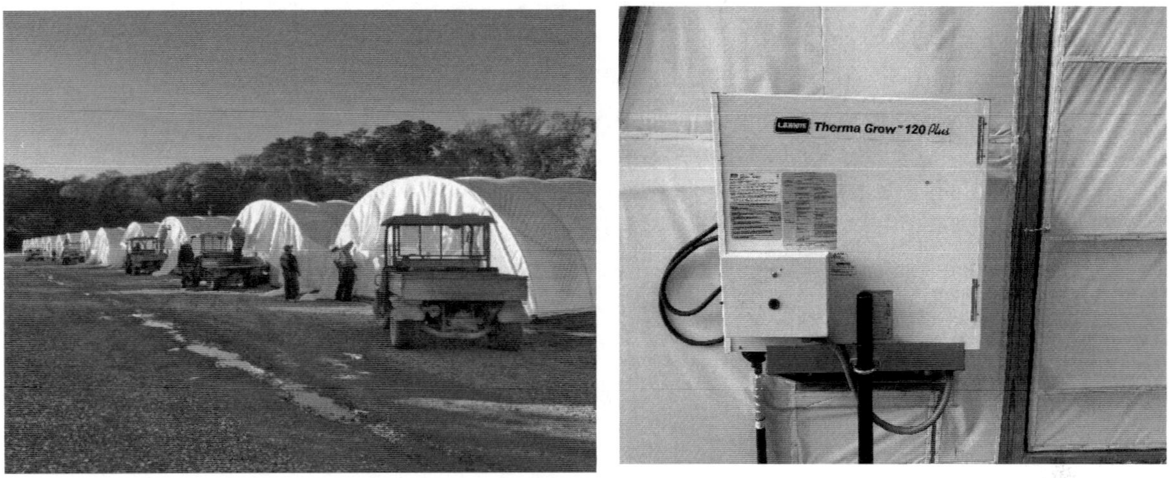

Figure 22. (left) Covering greenhouses with 50% opacity plastic for winter protection, and (right) outdoor mounted, direct fire, forced air heat for high efficency heating.

WEED MANAGEMENT

Preventative sanitation practices greatly reduce not only disease organisms but also weed seeds. Prior to filling each propagation house all debris is removed with a backpack blower. Next the house is sprayed with a disinfectant. Finally, a herbicide application is applied for weed prevention. Marengo® (a.i. indaziflam) or Sureguard® (a.i. flumioxazin) can be tank mixed with Roundup Pro® (a.i. glyphosate) and applied to the floor of an empty greenhouse (Fig. 23). Always review labels for instructions, precautions, and restrictions for all pesticides before application. The house is allowed to ventilate and dry for 24 hours before filling. The floor of each house is covered by a woven nylon ground fabric to prevent weeds as well. In addition, to sanitizing each house prior to filling with cuttings, either new or steam sterilized cell inserts are used (Fig. 24). Untreated used propagation trays are a potential source of both weed seeds and plant disease organisms.

Currently, Ronstar® G (ai. oxadiazon) by Bayer is the only preemergent herbicide labeled for use in propagation. It may not be used in an enclosed greenhouse. Therefore, it is applied after the overwintering plastic film is removed in the spring. Regular scouting for weeds, followed by timely hand weeding is necessary to manage weeds effectively.

Figure 23. Interior view of clean and sanitized propagation house.

Figure 24. New 18-cell and 21-cell tray inserts.

DISBUDDING

As cuttings are collected, all flower buds are removed to direct stored carbohydrates towards root development instead of flowering and seed formation. This also prevents botrytis blight on open flowers in the mist bed. Another round of disbudding is performed after cuttings have rooted and before the first flush of growth in the spring.

Figure 25. (left) Disbudding flower buds on rooted cuttings, (right) removed flower buds.

PRUNING

After the first flush of growth in the spring, liners are power sheared to promote branching. Blocks of liners are arranged in the house to accommodate a rolling gas-powered pruning machine. The overhead irrigation design allows for unobstructed access for the pruning machine and workers. All debris is blown and collected in a bag attachment on the pruning machine (Fig 26).

Figure 26. Gas powered sickle bar pruning machine with blower and bag attachment for uniform pruning and debris collection.

FINISHED LINERS

By early summer the following year, fully rooted branched liners are ready to shift up to either one-gal or three-gal containers (Figs. 27 and 28).

Figure 27. (left) Finished liners ready to pot into one-gal or three-gal containers, and (right) finished liners loaded onto trailers for transport to potting facility.

Figure 28. Well branched healthy camellia root systems.

Keeping Ornamental Winners and Losers in the Record Texas Freeze of February 15-21, 2021

David Creech, Lais Machado and David Kulhavy

SFA Gardens, Arthur Temple College of Forestry and Agriculture,
PO Box 13000, Stephen F. Austin State University, Nacogdoches, Texas 75962

dcreech@sfasu.edu

Keywords: Hundred-year freeze, tome, low-temperature - woody ornamental plant stress tolerance, plant evaluations

Summary

There was a record 100-year freeze in Texas from 15-21 February 2021. Temperatures dropped as low as -20°C (-4°F). For landscape plants, going from Zone 8b to a 5 was a bit much. Besides the low temperatures, the heavy ice and snow load further stressed plants in the Pineywoods of East Texas. We are developing a tome that describes the immediate and long-term impact of winter storm Uri on the Texas landscape. Recording a list of plants that thrived, survived or died would be useful to future landscape planners. While the tolerance of common plants was evaluated, the focus was on rarely encountered ornamentals. Stephen F. Austin Gardens (SFA) Gardens is a perfect platform to deliver freeze data - because it is a collector's garden of exotic plants. Hundreds of new plants are added to the landscape each year, which is a perfect crucible to test a wide variety of ornamentals exposed to extreme temperatures. The focus of this paper is limited to a few select genera, particularly those with adequate numbers for evaluation at SFA Gardens.

INTRODUCTION

In the last forty years, three freeze events stand out in Texas; December 1983, December 1989 and February 2021 (Fig.1). The most recent event, winter storm Uri, arrived in Texas in mid-February and every county in Texas fell under a freeze alert. Besides the human pain and billions in infrastructure losses, the winter storm emergency left a mark on the Texas landscape that will be long in the healing. The low temperatures broke records across the state. Nacogdoches is typically considered Zone 8B. Citizens were stunned when temperatures dropped to -20°C (-4°F) on February 16, 2021. City and residential water lines broke, the electric grid failed, and it was obvious Texas wasn't quite poised for record cold. For landscapes, going from Zone 8b to a 5 was a bit much. If it wasn't the cold, it was the heavy ice and snow load in much of the Pineywoods that proved too much. Patriarch pines, oaks, sweetgums and elms all suffered limb damage or total collapse.

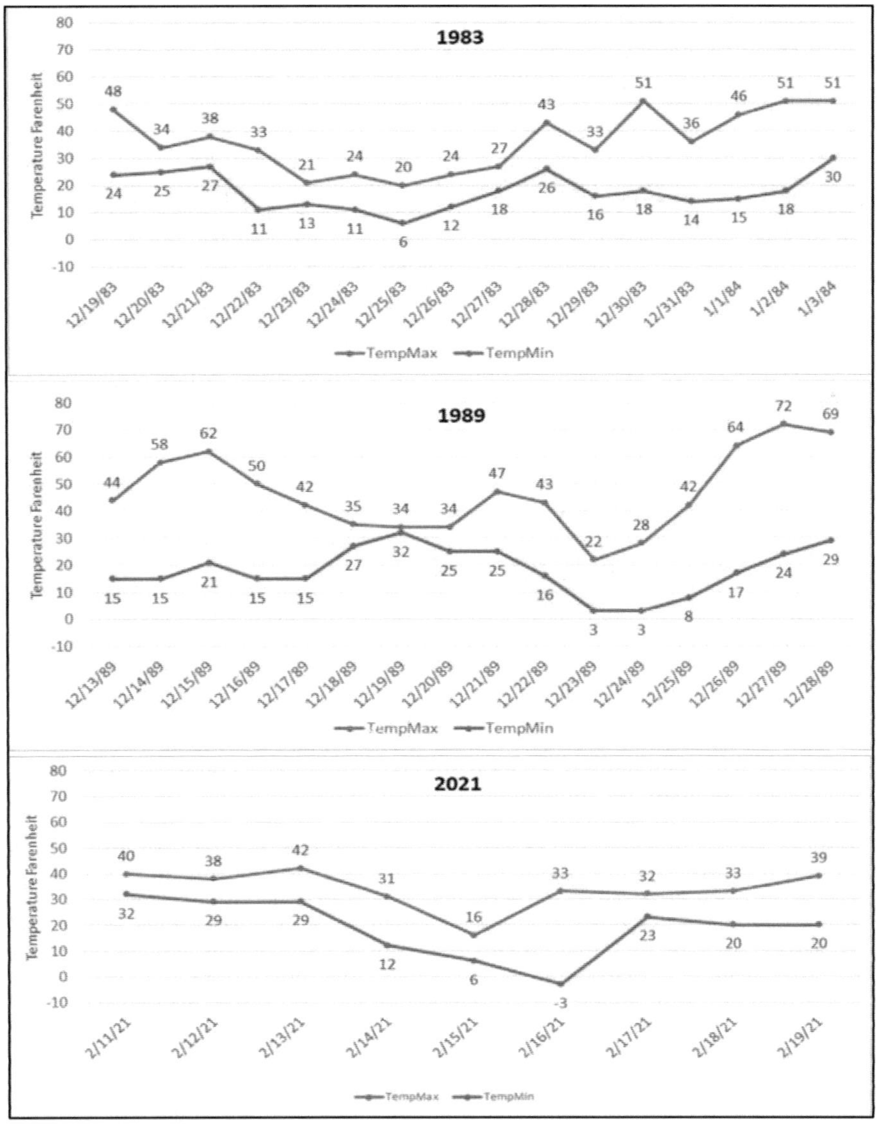

Figure 1. Temperatures encountered with three recent mega freezes, Nacogdoches, Texas

THE SFA GARDENS

For a background, SFA Gardens comprises 128 acres (58 ha) of on-campus property at Stephen F. Austin State University (SFA), Nacogdoches, Texas. SFA Gardens is the umbrella organization responsible for the activities, growth and development of five main theme gardens. Representing the oldest plantings, the 10-acre (4.5 ha) SFA Mast Arboretum was initiated in 1985, was dedicated in 1997, and includes the horticulture facility of the Agriculture Department. Second, the Ruby M. Mize Azalea garden was dedicated in April, 2000, and is an 8-acre (3.2 ha) garden of primarily azaleas, camellias, Japanese maples and an assortment of rarely encountered species and varieties. Third, the 42-acre (19 ha) Pineywoods Native Plant Center (PNPC) was dedicated by Lady Bird Johnson in April 2000. The SFA's Recreational Trail and Gardens was dedicated in March 2010 and comprises 68-acre (31 ha) acres of mostly undisturbed forest and includes the Gayla Mize Garden, an 8-acre (3.2 ha) garden that features woody ornamentals primarily. SFA Gardens is a collectors garden and features a wide diversity of species, varieties and genotypes: https://dcreech-site.com/2020/04/13/plant-glossary/.

Past record freezes in Texas

The February 14-17, 1895 snowstorm is still referred to as the Valentine's Day freeze, an event known for record snowfall on the Texas coast. Galveston reported snowfall over 15" with Houston, Orange, Stafford, and Columbus all reporting twenty inches. Even Brownsville at the southern tip of Texas received five inches and the huge "winter garden" vegetable industry was destroyed. To add to the wound, only a few years later, one of the worst winter storms ever in Texas struck Feb. 11–13, 1899. The entire state was impacted and newspapers then described it as the worst freeze ever known in the state. To this day, 1899 holds the record low for many Texas locations. There are other epic freezes in Texas history, of course. My Dad spoke of the 1929 freeze when ponds froze and it was bitterly cold for weeks. Yes, 1947 and 1951 brought serious low temperatures and 1960 brought record snowfalls. 2011 had a single digit cold snap and in January 2018, Nacogdoches dipped to 10°F for two nights in a row. However, in more recent history, there are two mega events that stand out. The December 1983 freeze event had statewide impact and lasted over two weeks. Six years later, the December 1989 freeze lasted two weeks with lows in the single digits and damage was everywhere. Ponds froze over, cattle and crops suffered and the zonal denial of the 1980s came to an end. It has been over thirty years since a really big freeze headline made the news. For many nurserymen and landscapers those events are only distant memories. While the February 2021 freeze lasted only a week, the record lows meant one thing. Texas has a brand-new benchmark for cold (Figure 1).

Objective

In the spring, a small group of horticulturists began a line of discussion that quickly concluded there should be an collaborative effort to gather freeze damage ratings for a wide range of ornamentals. After all, this was a 100-year freeze. We felt it would be prudent to put together a tome, one that describes the immediate and long-term impact of winter storm Uri on the landscape in Texas. Recording a list of plants that thrived, survived or died would be useful to future landscape planners. While the common commodities would be recorded, the focus would be on ornamentals rarely encountered. SFA Gardens is a perfect platform to deliver interesting freeze data simply because it's a collector's garden. Hundreds of new plants are added to the landscape each year, the perfect crucible to test a wide variety of ornamentals in a freeze event. For the purposes of this paper, the focus is limited to a few select genera, particularly those with good numbers at SFA Gardens.

METHODS

For SFA Gardens, the decision was made not to prune any landscape plants after the freeze until they showed the true impact of the winter storm. Somewhat coincidentally, a kiwifruit adaptation study underway happened to have six locations with dataloggers at co-operator locations and that data was captured (Fig.2). Galveston was not in the kiwifruit study but is added here to provide a southern coastal location. The graph is a combination of datalogger and available NOAA data. Dr. Mengmeng Gu, TAMU Agrilife Extension Specialist, Adam Black, premier plantsman, and a gathering of like-minded souls are accumulating the treasure trove of data available. A simple damage rating system was created by Dr. Gu. Basically, we're recording what thrived, survived or died. Together, we intend to build a statewide tome on how Texas landscape plants fared after winter storm Uri left the scene. The damage rating scale is rather simple:

At its most basic, this project will identify the location, genus, species, variety, damage rating and comments. In the midst of death and destruction, there's data. For the botanical garden community, this is an opportunity to create a reference point document for characterizing ornamentals for Texas with freeze tolerance in mind. There is nothing like a record breaker to define the field.

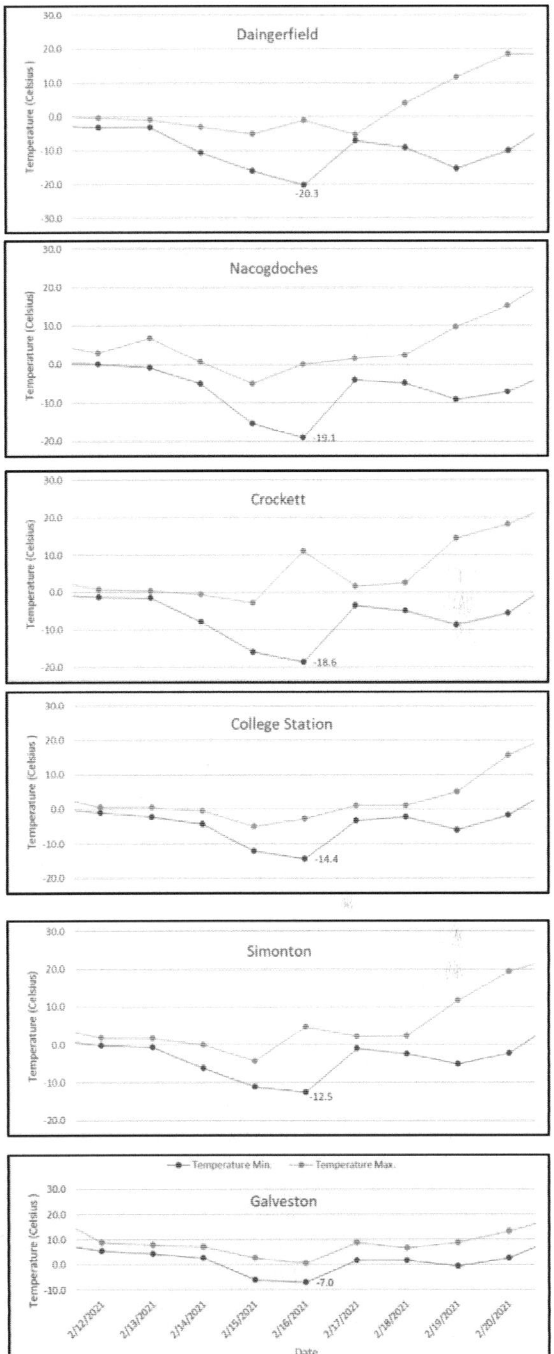

Figure 2. Maximum-minimum temperatures encountered in select Texas locations, 12-20 February 2021.

Table 1. Freeze damage rating system for woody plants

1: No damage
2: Minor foliar damage/partial defoliation, buds/stem survive
3: Near total foliar damage/defoliation, buds/stem push new growth
4: Outer branches dead, inner branches /main stem survive, likely to recover in 1-2 seasons without aesthetic disfigurement.
5: Major branches/main trunk damage, buds break usually from trunk, may have permanent aesthetic disfigurement
6: Total death

FREEZE IMPACT ON A FEW SELECT GENERA AT SFA GARDENS

Abelia – 16 cultivars, no damage – *A. chinensis*, no damage.

Acer species – The SFA Gardens Japanese maple collection is one of best in south and there's a good representation of rarely encountered Asian species. In general, most of the Acers suffered zero damage. Over 300 Japanese maples appeared to have emerged unscathed. However, the evergreen maples including *A. fabri*, *A. cinnamomifolium* and *A. oblongum* generally rated a 4 or 5 on the damage scale and are recovering. *Acer saccharum* ssp. *skutchi*, the Mexico mountain sugar maple suffered very little damage.

Actinidia – SFA Gardens and Texas A&M Agrilife have cooperated on a kiwifruit evaluation project for a number of years. For the most part, golden kiwifruit survived the freeze better than green, and young plants fared worse than older vines. A trunk protection study happened to be in place with temperature dataloggers and the conclusion was little to no benefit.

Berberis – mostly *B. thunbergii* varieties, no damage

Callicarpa – varieties and genotypes of *C. americana* suffered no damage. *C. rubella* and *C. dichotoma* damaged trunks and branches. *C. salicifolia* and *C. longissimia* froze to ground but both recovered.

Camellia – 200 plus cultivars with a wide range of damage ratings. Most survived though many badly damaged. 'Frank Hauser, a favorite here, was killed outright in a number of locations. 'Yuletide' branches and tops died back on some, on others less. For many Camellia species, it was common to have the top alive with unthrifty new growth with considerable sprouting from base and lower trunk and branches. Many straight Asian species died to near ground. *C. yuhsienensis* fared well.

Conifers – In general, good survivability over a wide range of genera including *Taxus, Cephalotaxus, Thuja, Thujopsis, Cunninghammia,* and *Juniperus*. Some damage on our three *Keteleeria* species and some nomenclature debate on our collection. A large *K. evelyniana* was killed back to trunk and a few major banches. A very large *Araucaria araucana* var. *angustifolia* (40' survived with some damage and new growth sprouting from trunk and the crown appears unaffected. *Cunninghammia unicanaliculata* (botanically challenged as a subspecies of *C. lanceolata*), weathered severe ice load and rebounded to good form without damage.

Gardenia – wide collection of varieties, froze to ground or near ground and recovered. 'Wispering Pines' was unaffected

Hydrangea – *H. quercifolia* and *H. paniculata* were unaffected. All *H. macrophylla* varieties froze to ground but returned vigorously. *Dichroa* survived from under snow cover.

Ilex – a large holly collection, unaffected for the most part. *I. rotunda* damaged. *I. vietnamensis* froze back.

Illicium – extensive collection. All native derived varieties seem to survive well, even the variegated and golden foliage clones. Surprisingly, *I. mexicanum* was unaffected. *I. anisatum* damaged. *I. verum* killed.

Lagerstroemia – 136 varieties, good survival but some varieties showing dieback and unthrifty growth, verdict not in.

Lauraceae – a record large *Cinammomum chekiangensis* was unaffected, a surprise. *Phoebe shearei* killed. *Phoebe chekiangensis* froze to near ground.

Loropetalum – a surprise, with major damage on a wide range of varieties, most to ground.

Magnolia – an extensive collection of varieties. *M. grandiflora*, *M. acuminata*, *M. pyramidata*, *M. virginiana*, and *M. macrophylla*, no freeze damage. However, some damage from snow/ice load. Many Asian magnolias suffered. The two banana shrubs, *M. figo* and *M. skinneriana* damaged, with *M. figo* frozen to ground. Surprisingly, a *Parakmeria yunnanensis* was unaffected.

Osmanthus – an extensive collection of *O. fragrans*. Most survived well. 'Fudingzhu' and 'Apricot echo' damaged but 'Aurantiacus' was not. Three variegated forms damaged but recovered from low in the shrub. *Osmanthus yunnanensis* taken to ground.

Pittosporum – all *P. tobira* varieties froze to ground but are returning. Both the green and variegated *P. heterophylla* froze to ground, sprouting from base and from underground roots, an aggravation. Some rarely encountered Asian Pittosporum species all froze to ground but have returned from base.

Podocarpus – the collection of varieties at SFA Gardens varied from major damage to little.

Quercus – an extensive collection of species. Damage to post oaks and live oaks in the region, but quite random. Some trees affected, other not. Most Mexico oaks in our collection survived in the landscape and in containers. Exceptions included *Q. germana* which suffered limb die back and unthrifty growth. *Q. tarahumara* froze back to main trunk and some side limbs. *Q. insignis* froze to near ground but has returned. *Q. rysophylla*, *Q. polymorpha*, *Q. canbyi*, three somewhat common in the Texas trade, all survived. A very large *Q. acutissima* died with no attempt at resprouting.

Raphiolepis – mainly *R. indica* varieties, most froze to ground. Indian hawthorns are a commodity in Texas landscapes and were badly damaged or killed all the way into Houston. *R. umbellata* survived with minor damage.

Rhododendron – With four hundred azalea varieties, selections or genotypes represented in SFA Gardens, Sherry Randall and Barbara Stump, both with long term involvement in the Azalea Society of America and the Texas chapter, made on-the ground evaluation in late spring. Esssentially no damage on native deciduous azaleas, Aromi hybrids, and other genetics in that arena. On *R. indicas*, it was typical to see alive but unthrifty tops with sprouting from base of plants. Encores in general were unaffected. 'Koromo shikibu', a signature azalea at SFA Gardens was unaffected. Badly damaged varieties were cut to a few feet above ground, fertilized and they have rebounded. Tables 1 and 2 present an example of the database used, sorted alphabetically by variety and by damage rating.

Schima – In the Theaceae, several species are now lumped into *S. wallichi.* Large tree at SFA Gardens that came to us as *S. superba* has damaged outer limbs, returned from trunk. Large *S. remotoseratta* died to ground but returned from base.

Styrax – The snowbells did well here. *Styrax japonica* varietries took the cold in stride, as did other Asian species, many rarely encountered. For example, *Styrax tonkinensis* was unscathed. A very large *Styrax formosanus* var. *formosanus* was killed to ground but vigorously sprouted from low on the trunks and from the root system. A large *Huodendendron tibeticum* (never flowered but grew well) was killed outright.

Taxodium – very large collection representing varieties and selection material of bald cypress, pond cypress, Monezuma cypress and the bald x Montezuma hybrids from the Nanjing Botanical Gardens Taxodium Breeding and Improvement program. No damage. This was a critical test of pure Montezuma genetics involving southern Mexico genotypes.

Ulmus parvifolia – Most *Ulmus* species were unaffected. However, in Texas, some large *U. parvifolia* trees were severely damaged with major limbs and trunk cracks.

CONCLUSIONS

Evaluating woody ornamentals for tolerance to a hard freeze event is more complicated than we originally envisioned. A few conclusions at this point can be made:

1. Patience is the rule. The impact of a freeze on a woody ornamental can take years to run its course. We have observed trees appear only modestly affected to observe them collapse.
2. With six inches of snow cover, many plants were protected and rebounded from below the snow line. A similar freeze without snow cover would have different results.
3. There's considerable variation in the data when multiple plants are involved. Whether seedlings or clones, there was obvious plant to plant variation. Assessing a variety's freeze tolerance on only a few plants may not be valid.
4. Numerous commodities need to be reconsidered. Loropetalum was introduced after the 1989 freeze, planted extensively in Texas and was badly damaged by the February 2021 freeze. While the species generally resprouted from the base, robust sprouts from the root system are a maintenance aggravation.
5. The final document will be available in .pdf format and placed on the web for future reference.

Cutting Propagation of *Magnolia grandiflora*

John Davy

Panhandle Growers, Inc., 5907 Southridge Road, Milton, FL 32570 USA

johnmdavy@aol.com

Keywords: Indolebutyric acid (IBA), naphthaleneacetic acid (NAA), semi-hardwood cuttings, Southern Magnolia.

Summary

In Northwest Florida (Zone 8B), propagation of *Magnolia grandiflora* with semi-hardwood cuttings is done from 15 August to 30 November (first frost), depending on the cultivar. Soil mix and flats are drenched with the fungicide, Subdue®, before sticking. Cuttings are 10-15 cm (4-6 in.) in length, with bottom 1-2 leaves removed. Cuttings are scored 2.5 cm (1-in.) on one side, then quick-dipped for 30-sec in 10,000 ppm indole-butyric acid with potassium salt (K-IBA). 'Brackens Brown Beauty' is also treated with 500 ppm naphthaleneacetic acid (NAA). Avoiding over-watering cuttings is critical. Foliar application of Peters® N-P-K is applied to cuttings at low rates after callus develops. After adventitious roots appear, cuttings are drenched with the fungicide, Safari®. Rooting success rates of cultivars are as follows: 'Claudia Wannamaker' 95-100%, 'DD Blanchard' 95-100%, 'Kay Parris' 80-95%, 'Bracken's Brown Beauty' 70-85%, 'Little Gem' 60-95%, 'Opal Beach'80-90%, and 'Seagrove' 80-95%.

INTRODUCTION

Stock plant maintenance

Propagation of *Magnolia grandiflora* with semi-hardwood cuttings in Northwest Florida (Zone 8B) begins with proper stock plant maintenance (Fig. 1). Most producers grow stock blocks to generate cutting material; some use production plants, while others use a combination of both. At Panhandle Growers, the in-ground field production plants serve as stock (Fig. 1). Stock plants should be growing vigorously, free from insects and diseases and fully hydrated at the time of the cuttings.

Figure 1. Cutting propagation of select *Magnolia grandiflora* cultivars. (left) Classic white aromatic magnolia bloom, (center) stock blocks from which semi-hardwood cuttings will be taken, and (right) a finished, field-grown *M. grandifora* being prepped for harvest as a balled & burlap, tree.

Condition of the cutting wood

This is perhaps the most important ingredient in success. Timing is everything. Semi-hardwood cuttings from the final flush of growth are best. Stick dates are dictated by the timing of the stock plants moving toward winter dormancy (Table 1).

Table 1. The average dates for perfect cuttings from field production in Northwest Florida (Zone 8b).

Cultivar	Range of rooting success
Bracken's Brown Beauty	Aug 15-Sept 7
DD Blanchard	Aug 20-Sept 15
Kay Parris	Aug 30-Sept 15
Claudia Wannamaker	Sept 1-Sept 15
Little Gem	Sept 1-Nov 30 (first frost)

Poorer results have occurred with cuttings taken later in the season from field plants that received late shape pruning and plants grown in containers. Late season rooting percentages can be increased with bottom heat.

Cutting preparation

Cuttings are taken early in the day, typically before 10am, bagged, hydrated with ice and placed in the shade. Cuttings are processed and stuck within one hour. Cuttings are 10-15 cm (4-6 in.) in length with 4 to 6 leaf nodes taken. The cutting base is cut at a slight angle with sharp, sterilized pruners. Larger diameter cuttings are preferred. The color of the foliage helps to determine condition of the wood. Cuttings with terminal growth buds are preferred.

Propagation media

Various propagation mixes and media have been used. At Panhandle Growers, Inc. (PGI) we use 60% aged fine particle pine bark and 40% perlite. A minor element package is added as is dolomitic pelletized lime to bring the pH to 6.5. No N-P-K fertilizer is added to the mix. Propagation is done in multi-celled flats (Fig. 2). Each propagation flat contains 18 square cups, which are 9 cm (3.5-in.). They are filled and pre-staged on the benches. Soil mix and flats are drenched with the fungicide, Subdue®, before sticking. It is important to start with a clean, dry house.

Figure 2. Hard-wood cutting propagation of select *Magnolia grandiflora* cultivars under intermittent mist. Each cutting is stuck in an individual cell of a multi-cell propagation flat.

Preparation of the hard-wood cuttings

The bottom 1 or 2 leaves are removed from cuttings. Cuttings are scored 2.5 cm (1-in.) on one side, then quick-dipped for 30-sec in 10,000 ppm indolebutyric acid with potassium salt (K-IBA). 'Brackens Brown Beauty' is also treated with 500 ppm naphthaleneacetic acid (NAA). Cuttings are stuck no more than 2.5 cm (1-in.) deep in the media.

Water management

Water management is critical for success. Many propagators over water southern magnolia cuttings. The time clock settings used at Panhandle Growers vary depending on rooting stage during propagation (Table 3). Weather and callous development dictate change in program. After 4 to 6 weeks cuttings are watered manually through the mist lines or spot watered as needed (Fig. 3).

Table 2. Mist settings during southern magnolia cutting propagation.

Program setting	Mist timing
Initial	Start time 9am (early season) Stop time 3:30
Program 1: first week	10 second duration with 5-minute delay
Program 2: second week	15 second duration with 10-minute delay
Program 3: third week	30 second duration with 25-minute delay
Program 4: fourth week	60 second duration with 1 hour delay

Figure 3. Rooting of *Magnolia grandiflora* cuttings under intermittent mist.

Cultivar rooting Success

Rooting success in southern magnolia dependent on cultivar being rooted (Table 1), stock plant treatment, time of year, substrate and cutting environment.

Table 1. Rooting success in selected cultivars of southern magnolia.

Cultivar	Range of rooting success
Claudia Wannamaker	95-100%
DD Blanchard	95-100%
Kay Parris	80-95%
Bracken's Brown Beauty	70-85%
Little Gem	60-95%
Opal Beach	80-90%
Seagrove	80-95%
Teddy Bear	Unknown

Long term care

Foliar application of Peters® N-P-K at low rates is applied to cuttings after callus develops. Every 3-4 weeks cuttings are drenched with a low rate of N-P-K through the winter months. After adventitious roots appear, cuttings are drenched with the fungicide, Safari® (Fig. 4). This treatment is done late in the day to reduce environmental impacts. Typically, no other fungicides or insecticides are needed.

Practicing cleanliness is essential by keeping the floors clean of fallen leaves, weeds, and loose soil mix. The propagation houses are walked weekly to remove any fallen leaves – and keep clean. Once rooted, the propagation house is kept cool. Nighttime temperatures are allowed to drop to 2°C (35°F) with no damage. Exhaust fans start at 29°C (85°F).

Figure 4. Adventitious rooting of *Magnolia grandiflora* stem cuttings from (left) a single adventitious root (arrow) to (middle, right) multiple roots along the stem cuttings.

Moving out rooted liners

At the last chance of freezing temperatures (typically mid-April) the flats are moved to a 30% shade house and given a low rate of slow release 17-5-11 N-P-K with minors (Fig. 5).

Liners are graded and shifted to a Rootmaker® 3-gal containers by mid-May (Fig. 6). Most cultivars will be suitable for shifting to a larger container or for field planting by November, except for 'Little Gem' (Fig. 7).

Figure 5. Rooted southern magnolia plants ready for transplanting.

Figure 6. Containerized liner production of rooted *Magnolia grandiflora* produced from cuttings.

Connecting Veterans to Horticulture

Robert Elliot[1] and Lis Meyer[2]

[1]Veteran's Farm of North Carolina and Soldier to Agriculture Program, North Carolina State University, Raleigh, North Carolina 27695, USA; [2] Department of Horticultural Science, North Carolina State University, Raleigh, North Carolina 27695, USA.

vetfarmofnc@gmail.com; lis_meyer@ncsu.edu

Keywords: Career Skills Program (CSP), Soldier to Agriculture Program (STAG), Transition Assistance Program (TAPS), leadership, nursery industry profession, superior problem-solving skills, work ethic, Veteran's Farm of North Carolina, Inc. (VFNC), Veterans Agricultural Training and Education Program (VATEP)

Summary

Agriculture, the green industry and horticulture can help returning veterans integrate back into civilian life with productive careers. Likewise, the human talent and skillsets that veterans offer can be a great employee resource for the green industry. After 15-years of service, veteran Robert Elliott returned to his family farm and developed a profitable small farming operation. Suicide is the second leading cause of death in the US military. His success ("*Farming saved my life*") – the story of a veteran finding peace and a life worth living with real purpose - brought a flood of veterans to Robert's doorstep to learn how to become farmers. Robert began working with veterans across the country and developing programs for those interested in agriculture. He also continued with his own education, pursuing a B.S. in Biological and Agricultural Engineering with a minor in Horticultural Science at North Carolina State University (NCSU). Lis Meyer of NCSU not only introduced him to the possibilities of plant propagation as a science, but also career opportunities in the nursery industry. One of the initial programs Robert started in his efforts to connect veterans with agriculture, which first involved Lis - was the Soldier to Agriculture Program (STAG) at NCSU. Robert went on to also start the Veterans Farm of North Carolina (VFNC), which provides consultation and training to veterans and transitioning military on agricultural production methods of all scales and sizes. Its newest program is a six-month, hands-on farm training program known as the Veterans Agricultural Training and Education Program (VATEP).

INTRODUCTION

In North Carolina, agriculture (including the nursery industry) is the number one source of income for the state, followed closely by the United State military. While historically there was much overlap between the farming and military populations not only in NC, but also the US as a whole, these two professions began to separate a great deal particularly following the passage of the GI Bill in 1955.

Robert Elliott started out following a similar path as that of many in the U.S. armed forces—he grew up on a large, commodity crop and livestock farm in Franklin County during the late 1980's and 1990's. Seeing his family struggle to keep their farm afloat, Robert had no desire to continue in agriculture initially and at 18 years old joined the U.S. Marines.

Robert was honorably discharged due to an injury after five years in the Marines, after which he worked with his same unit as a DoD contractor for an additional ten years. Following being laid off from his contractor position, Robert's story diverged from the typical narrative of separation between military and agriculture as he found himself back in NC on his family farm. He turned the farm into a profitable small farming operation based on niche pork before his story began to attract media attention. Robert was interviewed about how taking to the land and farming helped him avoid the post-military suicide statistic that many veterans face today.

When a veteran returns home, they lose their entire support network, and it is one of the most challenging things they face that can ultimately lead to a hopeless view on life. A quote from Robert's interviews hit home to many veterans that read his story: "Farming saved my life." The stories of a veteran finding peace and a life worth living with real purpose brought a flood of veterans to Robert's doorstep to learn how to become farmers themselves.

He began working with veterans across the country and developing programs for those interested in agriculture. He also continued his own education, pursuing a B.S. in Biological and Agricultural Engineering with a minor in Horticultural Science at North Carolina State University. In 2019, Robert enrolled in a Plant Propagation course taught by Lis Meyer, a senior lecturer in the Department of Horticultural Science. This class introduced him to the possibilities not only of plant propagation as a science but of the nursery industry.

Lis Meyer, in turn, had seen a steady increase in veteran enrollments in her classes and programs since she first began teaching in 2010. At any given point during the past five years, at least a quarter of the students enrolled in the Undergraduate Certificate Program in General Horticulture which she coordinated were veterans or veteran spouses. Having observed the strong work ethic, superior problem-solving skills, and enthusiasm for learning that these veteran students brought to the classroom, Lis was happy to accept Robert's invitation to teach plant propagation sessions as part of the Soldier to Agriculture Program which he had started at Fort Bragg through the NC State Agricultural Institute. This partnership opens the door to a whole new world of possibilities both for the veteran students eager to learn more about horticulture and for the field of nursery production as we search for the next generation of leaders for our industry.

Soldier to Agriculture Program (STAG)

One of the initial programs Robert started in his efforts to connect veterans with agriculture, and the one in which Lis first became involved, was the Soldier to Agriculture Program (STAG) at North Carolina State University (NCSU). While developing a connection between the military and NC State's Agricultural Institute (AGI), a request from the Fort Bragg Transition Assistance Program (TAPS) officer was made to Robert and Dr. Elizabeth Wilson, the director of AGI. Many of the 750 soldiers per month that were exiting the Army into the civilian world from Fort Bragg were looking for a program that would help them learn how to start a farm. TAPS is a program within the Army that exists to help soldiers find a smooth transition into the civilian world.

STAG is placed within the Career Skills Program (CSP) of TAPS. CSPs is a relatively new program developed by the Department of Army to reduce the unemployment benefits paid to veterans once they exit the military. Research showed that soldiers were not prepared adequately nor timely enough to acquire a job immediately upon exit. The CSP model offers corporations from many industries an opportunity to come to Fort Bragg, acquire office and classroom space, and gain direct access to transitioning military personnel that could be released from their unit temporarily to be recruited by these companies and train for jobs while in their last 180 days of active-duty service. Upon completion of the CSP of the soldier's choice, they would have a job waiting on them once they fulfilled their contract term in the Army. STAG became an extremely popular CSP with Fort Bragg, and now, across all branches of service. STAG is regarded by TAP staff as, "*The most sought-after transition program on Fort Bragg.*" To date, there have been 26 total cohorts with 290 graduates - and there is a waiting list of over 490 interested in attending the class.

STAG is typically an in-person class that runs Monday through Friday for six weeks from 09:30-13:00. Since March of 2020, the course has been completely online due to COVID with the only exception being optional, COVID-compliant tours on farmer veteran operations in the area. Lis has taught one to two-day plant propagation sessions as a guest instructor with STAG since 2019. During these sessions, STAG students learn and get hands-on experience with the basics of greenhouse mist systems, cutting propagation, grafting, bulb scaling, and much more (Fig. 1).

Figure 1. Students learning how to graft ginkgoes (*Ginkgo biloba*) in the Soldier to Agriculture Program (STAG) at Ft. Bragg, North Carolina.

They are also introduced to various facets of the nursery industry, including professional organizations like IPPS (Figs. 2, 3, and 4)

Figure 2. A veteran and former Soldier to Agriculture Program (STAG) student with his family at a holiday houseplant sale held to support the program.

Figure 3. Lis Meyer, Robert Elliott, Samantha Manning, and other Soldier to Agriculture Program (STAG) participants at a holiday houseplant sale held to raise funding for the program.

Figure 4. Soldier to Agriculture Program (STAG) students visiting the Fox Greenhouses at North Carolina State University.

Veteran's Farm of North Carolina (VFNC)

Robert also built the Veteran's Farm of NC, Inc. (VFNC), a 501(c)(3) non-profit organization dedicated to assisting veterans and military personnel/unit commands with consultation, training and education, equipment usage, and land acquisition (Figs. 5 and 6). Founded in 2014, VFNC began to bridge the gap between the military and agricultural communities. VFNC works through five main programming areas: recruitment, consultation, networking, education and training, and Infrastructure.

VFNC recruits interested veterans both in and out of the service and exposes them to all of the opportunities agriculture has to offer. Veterans learn that the agriculture industry has many desirable career opportunities. Recruiting also means educating a veteran before they start farming to develop the skill sets, have a plan, minimize risks - and the foundation to develop an economically successful business. The VFNC helps create best management practices (BMPs) through education based on past experiences and real knowledge - from experienced members of agriculture-related industries.

Figure 5. (top & bottom) Robert Elliott, Veteran's Farm of North Carolina (VFNC) and Soldier to Agriculture (STAG) Program.

VFNC provides consultation and training to veterans and transitioning military on agricultural production methods of all scales and sizes. Its newest program is a six-month, hands-on farm training program known as the Veterans Agricultural Training and Education Program (VATEP). In short, VATEP will have the primary models of agriculture that most farms engage in today - on a newly acquired 53-acre parcel of land in Cameron, NC. Participants in the program will have the opportunity to work with each production model such as beekeeping, livestock, greenhouse/horticulture, agronomy/field crops, pasture management, forestry, hydroponics, indoor food production, and much more. This program is partnered with Fayetteville Technical Community College and students will complete online coursework in a variety of subjects with hands-on, practicums to be completed on the farm (Fig. 7). Upon completion, participants show proficiency in all business requirements a farm requires by practicing those principles on the farm with guidance. Many more benefits for veterans will be available from VATEP and an extended program guide is available for those with interest in the program.

VFNC has built a network of over 500 farmer veterans and links them to critical resources in NC. VFNC believes that collaboration is the only way farmers can succeed. VFNC also assists veterans with acquiring lower cost land and equipment rental. VFNC accepts financial donations, and gifting of agricultural and general construction equipment, as well as infrastructural items (Fig. 8). These items are available for veterans to use in NC. We currently maintain an inventory of over $200,000 in equipment including tractors, implements, trailers, a truck, freezers, other equipment - and much more. These items are heavily utilized. This initiative is instrumental in assisting a farmer with their endeavors and sometimes can be an absolute game-changer for a farmer that cannot afford the tractor but stands to greatly multiply their production and revenue by using it for a few days.

VFNC's mission is, "To give veterans a new mission and America new farmers."

IPPS Involvement

In January 2020, the IPPS Southern Region of North American executive committee unanimously voted to extend free student memberships to those enrolled in the STAG program.

This decision was an excellent first step in bringing together two groups of people who have much to offer each other. We look forward to seeing this relationship grow and development in years to come, as IPPS members "Seek and Share" the valuable knowledge that they have with those veterans who wish to become our colleagues.

35-Years of Propagation in 30 Minutes – Tricks and Tips

Maarten van der Giessen

van der Giessen Nursery Inc., 4701 Seabrook Road, Wilmer, AL 36587 USA

maarten@bestliners.com

Keywords: Liner production, Semmes, Alabama nursery industry, propagation media, Phytotronics 1626 Clock® mist controller, side-veneer graft, softwood, semi-hardwood and hardwood cuttings, potassium salt-Indole-butyric acid (K-IBA)

Summary

Van der Giessen Nursery is a 20-ha (50-ac) wholesale liner and container growing operation with a wide pallet of woody ornamental plants. We produce over two million liners a year. Our timeline for propagation begins in January and ends in December. We have five keys to successful propagation: 1. use juvenile stock -if limited to field grown material, it is best to coppice the material and take the resulting flush; 2. proper nutrition and healthy stock are critical - a nutritionally-stressed cutting will never make a good liner; 3. know what growth stage is best to maximize rooting success: softwood, semi-hardwood, or hardwood cuttings; 4. know the optimal window of opportunity to take successful cuttings; and 5. if you have correctly managed 1-4, then optimize use of rooting hormones. Recommendations are given on propagation of select species.

INTRODUCTION

Van der Giessen Nursery is a 20 ha (50-ac) wholesale liner and container growing operation with a wide pallet of woody ornamental plants (Fig. 1). We produce over two million liners a year. The nursery began in 1990 upon the retirement of my father, Peter van der Giessen, as manager of Cottage Hill Nursery in Irvington, Alabama (Fig. 2). Two months after he "retired" dad called me to join him for lunch in Semmes, Alabama, a small farming community near

Mobile. Halfway through lunch he asked "What do you think about the old Lyons Nursery?" I told him it was a wreck, and had been abandoned for years. "Good: I bought it! Want to go to work!?" So, we began!

Figure 1. Van der Giessen Nursery is a 20-ha (50-ac) wholesale liner and container growing operation with a wide pallet of woody ornamental plants.

Figure 2. The late, Peter van der Giessen. He and son, Maarten, are the 1st father and son to be recognized as Fellows of the IPPS-Southern Region of North America!

Semmes, Alabama has been home to wholesale nurseries for over a hundred years. It has good soils, plenty of water, and access to rail; this attracted Kiyono and Sawada from Texas, and the Welch Brothers from Iowa to start businesses in Semmes during in the first decade of the 20th Century (Fig. 3). They spawned a nursery industry that thrives to this day. The friendship of the nurserymen in Semmes also makes it an ideal home for a wholesale nursery.

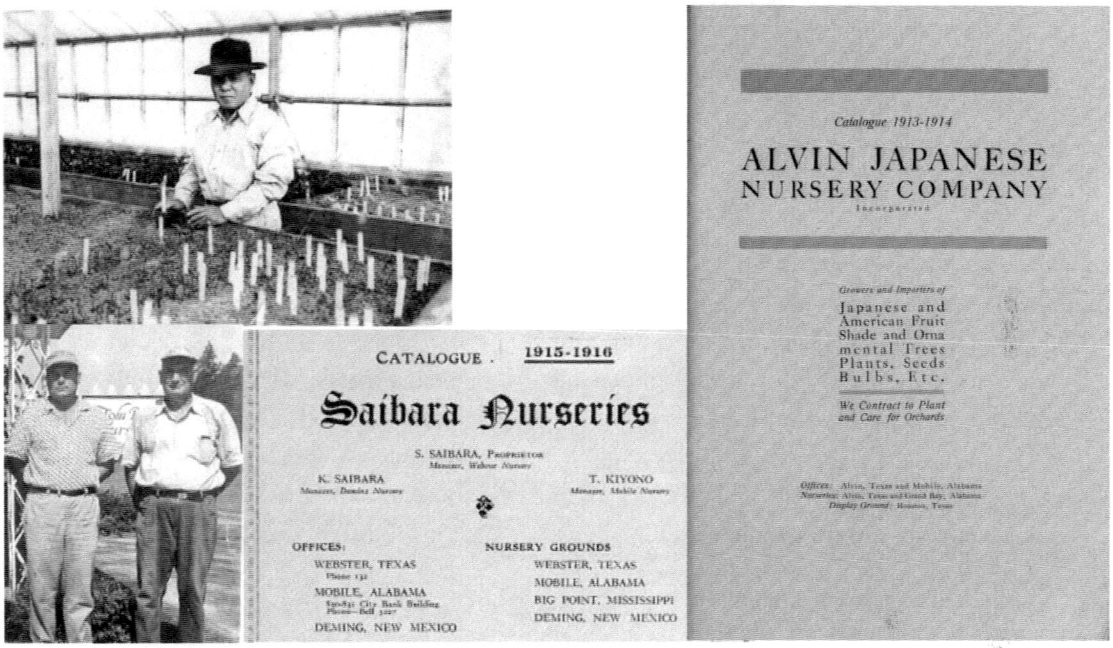

Figure 3. Early pioneers of the Semmes, Alabama nursery industry: Clockwise from top: Kosuku Sawada, Alvin Nursery catalog (Sawada), Saibara Nursery catalog (Kiyono), and Tom Dodd Jr. with Tom Dodd Sr.

Media mixes

Pine bark is still plentiful in our area. Our mix at van der Giessen is a 3 bark: 1 peat: 1 pine shavings - blended with lime, micronutrient, gypsum, and fertilizer amendments (Fig. 4). We like the pine shavings for propagation for two reasons. The shavings initially fluff the mix and provide improved drainage. As they break down, they act more like a peat, holding moisture and binding the media together.

Figure 4. Soil mixes. (left) The general soil mix of 3 bark: 1 peat: 1 pine shavings; and (right) for conifers, the mix is 3 bark: 1 peat: 1 pine shavings: 1 perlite.

Propagation structures

Our structures are typical 7 x 30 m (24 x 100 ft) Quonset houses, which are popular in the area (Fig. 5). The mist system is run by Phytotronic 1626® timers throughout the nursery. We prefer the Phytotronic clock for its simplicity. A good philosophy is: "If your employees cannot operate it, then maybe you should not use it either". It can be changed out in five minutes or less, and does not require a manual to operate. Our propagation sprinklers are either the Rain-bird MaxiBird® or Senninger Wobblers® (Fig. 6). Both are sturdy, inexpensive, and reliable. We produce liners in 40 cell 1801 flats. A house holds 1100 trays – and 44,000 liner plants.

Figure 5. A 7 x 30 m (24 x 100 ft) shaded Quonset house for propagation and rooted-liner production. Each propagation tray contains 40-cells 6 x 8-cm (2.3 x 3-in.)] with 1100 trays per house - and 44,000 liner plants.

Figure 6. (left) Phytotronics 1626 Clock® controller (arrow) can mist up to 6-zones https://www.phytotronics.com/product/nova-1626et-6-zone-controller/ . (right) Our sprinklers include the Rainbird Maxi Bird® at 3-gal per minute, and Senninger Mini-Wobblers®.

There is no substitute for personal attention in propagation. No amount of sophisticated gadgetry will replace a good propagator. As a safety precaution, we hang two-sided plastic flags on our clock boxes. One side is green, the other side is red. We check our timers at 8:00 and, if all is well, we hang the green flag on the clock box. We check again that the houses are cycling properly at 10:00 and 14:00. At 16:00 we make one final round and take off the flag. If during the day an employee needs to turn off a house, they are required to flip the flag from green to red. We are all human! I cannot tell you how many times in 35 years I have seen houses turned-off. Here at van der Giessen - turning off a clock without red flagging it - is a capital offense. That five-cent piece of plastic has saved us tens of thousands of dollars.

Our use of poly in propagation has changed over the years. We use 55% white poly January through March (Fig. 7). While most nurseries propagate conifers in open beds, we prefer the greater control of moisture and temperature from white poly in overwintering houses. From April through October, we propagate under 55% saran with no poly. This method evolved in response to 2005's hurricane season culminating in Hurricane Katrina, which destroyed 125 billion dollars' worth of property on the Gulf Coast. We saw five hurricanes impact our area that year (Fig. 8).

A crew of four men can strip a plastic covered house and secure the plastic in 20 minutes. For fifty greenhouses that means it takes 66-man hours just to disassemble. It takes the same amount of work to replace the poly when the storm has passed. In 2005, we calculated that we spent one entire month pulling and replacing poly. The labor and materials were incredibly expensive. After 2005, we decided to try propagating under shade cloth with no poly. A crew of two can strip shade cloth in 15 minutes, cutting our labor by more than half. To our pleasant surprise we found that not only could we propagate without poly, but in some instances the plants were healthier.

Figure 7. Our Quonset propagation house coverings alternate from (top, left) 55% saran with no poly from April through October; (top, right) white poly from January through March, and (bottom) clear poly with 55% saran from October through December.

Labor costs have increased by almost 70% in the past fifteen years. During the same period our pool of available labor has shrunk dramatically. Plant pricing in the industry has not kept pace with production costs. This has driven the industry to find ways to cut costs. At van der Giessen, we worked with Ellis Manufacturing to build a 15-bit drill-head for our potting machine capable of producing jumbo quarts (Fig. 9). We decided to move away from the traditional 1-gallon container for much of our production. A 190 m^2 (2000 ft^2) greenhouse will hold 15,000 quarts in flats versus 5000 1-gal pots. Additionally, we can box and palletize the quarts to ship with our liners.

This gives our customers an opportunity to step up new introductions directly to 3-gal containers at a competitive price with 1-gal pots.

Figure 8. On 24 October 2020, Hurricane Zeta destroyed eight Quonset houses.

Figure 9. (left) Ellis Manufacturing built a 15-bit drill-head for our potting machine capable of producing jumbo quarts. (right) We can box and palletize the quarts to ship with our liners.

Five keys to successful propagation

Our timeline for propagation begins in January and ends in December (Fig. 10).

We have five keys to successful propagation:

1. Use juvenile stock. If you are limited to field grown material it is best to coppice the material and take the resulting flush.
2. Proper nutrition and healthy stock are critical. A nutritionally-stressed cutting will never make a good liner.
3. Know what growth stage is best to maximize rooting success: softwood, semi-hardwood, or hardwood cuttings.
4. Know the optimal window of opportunity to take successful cuttings.
5. If you have managed 1-4, then optimize rooting hormone applications.

Figure 10. Proper seasonal timing of collecting cutting wood is critical in propagation.

Dividing conifers into easy and difficult classes

We divide our January/February conifers into two classes: easy and difficult (Fig. 11). *Chamaecyparis, Thuja, Juniperus horizontalis* and *J. conferta* are generally easy to root. On the other hand, *Juniperus chinensis* 'Blue Point', *J. procumbens* nana, and *J. virginiana* can be difficult.

We make sure to keep them separated from the faster rooting stock. A *Juniperus horizontalis* 'Blue Rug" will have rotted long before a *J. chinensis* 'Blue Point' has started. *Cryptomeria* is always propagated in a dedicated house for the same reason. The key to rooting tougher conifers is not to over-water. We syringe those houses four times a day.

Figure 11. Dividing conifers into easy- and difficult-to-propagate species.

Grafting of maples

In February, we graft *Acer palmatum* and *A. japonicum*. We bring the understock into a marginally heated house [2-4°C (35-40°F)] in January. We use a side-veneer graft with grafting tape. We do not bag. We have a 90% success rate. We graft onto our understock at the second or third node above the soil. This gives us the opportunity to re-graft onto the same stock if the initial graft does not take. The key is to keep the graft union dry. Since the plants are dormant, they do not require a lot of water. Once they begin to break dormancy, we hand water below the graft for three weeks until the graft has healed.

Figure 12. Grafting maples using the side veneer graft (right, arrow). Understock is brought into the grafting house in January.

Figure 13. (left) Bench grafting maples using a side veneer graft (right, arrow). The darker, smaller scion piece is held in the grafter's hand.

March: propagation of *Cephalotaxus*

In March we propagate *Cephalotaxus harringtonia* (Fig. 14). We have found that we get better response if the stock is about to break dormancy. *Cephalotaxus* benefits from a 5000 ppm K-IBA quick-dip. The key to *Cephalotaxus* is patience.

It is not uncommon for a crop to sit for three to four months before they begin to root in good percentages. We keep the house covered in white poly and syringe three or four times a day. We typically get 80% rooting. This is insufficient for liner production. Consequently, this crop is later transplanted into quarts for sales.

Figure 14. Propagation of *Cephalotaxus* in March.

April: A busy month! More propagation

April is the cruelest month. We are shipping like crazy and there are some plants that simply will not root in good percentages – unless you use the first growth flush for softwood cuttings (Fig. 15). *Acer japonica, Rhododendron*, and *Chionanthus retusus* are all propagated as softwood cuttings. I define a softwood cutting as a 10-15 cm (4-6 in.) cutting that can be bent double without breaking. If you bend it end to end and it snaps, we consider it semi-hardwood. If a native azalea is semi-hardwood, then it is too late; rooting percentages will drop by 50%. Additionally, we take deciduous Ilex species early. We get better rooting, and the rooted liner has more time to develop a flush of growth. As with native azaleas, an Ilex that roots but does not flush will have a difficult time breaking dormancy the following winter. We root lepidote and elepidote *Rhododendron* in April as well. They are another group that will sit for a long time under mist and then suddenly throw a pot full of roots. Be patient.

Figure 15. Propagation of deciduous azaleas using soft to semi-soft wood cuttings.

May: Propagating with semi-hardwood material

By May we are getting into the semi-hardwood material, which is the bulk of what we propagate cuttings with (Fig. 16). This is a good time to propagate *Ilex vomitoria* by cuttings. Yaupon can be taken either early spring of early fall successfully. The key seems to be that it does not like extremes of hot or cold. Some nurseries use up to 10,000 ppm K-IBA with yaupon cuttings, but my personal experience is to use no rooting hormone, or employ a mild (1250 ppm) quick-dip of K-IBA – for best rooting. Sodium salt IBA always burns the cuttings. Another plant that seems to respond well to propagation in May under saran is *Distylium*; we use 2500 ppm K-IBA with 95% success.

Figure 16. Semi-hardwood cuttings used to propagate (left) *Distylium* 'Spring Frost', (right) *Ilex vomitoria* 'Oscar Grey'. *Ilex vomitoria* and *Chaenomeles* require no rooting hormones, whereas *Berberis*, *Cliftonia* and *Distylium* are treated with 2500 ppm K-IBA.

June: Evergreen azalea season

June is evergreen azalea season (Fig. 17). We stick around 400,000 azalea cuttings a year. We no longer strip our cuttings nor cut the tips.

Figure 17. Semi-hardwood cuttings of evergreen azaleas are neither stripped nor quick-dipped with auxin.

July and August propagation

July and August are our best months for *Viburnum macrocephalum*, *V. tinus*, and *V. opulus*. The larger viburnums: *V. awabuki,* and *V. sandanqua* will root well - but overgrow a 40-cellpack before spring. Additionally, this is the best time to root *Myrica*, *Osmanthus*, and *Edgeworthia*. *Myrica* stuck early will become too leggy for spring sales. Later in the season it becomes difficult to root in good percentages.

September propagation

In September temperatures fall back into an average range of 27°C (80°C). It is a good time to propagate *Ilex vomitoria* (Fig. 18). I find that *Fothergilla*, *Pieris*, *Illicium*, and *Leucothoe* also prefer the moderate temperatures.

Our *Cleyera* seed is ripe in September. We discovered that rotting the seed in a mist house for three weeks after picking seed gives us much better germination rates; but it is a stinky job to screen and wash them. We sow them in flats with no fertilizer with a light covering of perlite.

Figure 18. Propagation of *Ilex vomitoria*.

October and November propagation

By October and November - we are running out of time. Still, many plants will root if cuttings are taken up to the end of November. *Loropetalum, Eleagnus, Buxus, Pittosporum, Michelia*, and *Podocarpus* will root if cuttings are taken by or before the end of November.

December propagation

By the beginning of December, we experience our first hard frosts. We can propagate easy-to-root conifers and a few more elipodote rhododendrons. Then we start a new season all over again in January.

The Does and Don'ts of Using Biologicals

Chris Hayes

BioWorks, Inc., 100 Rawson Rd., Ste. 205, Victor, NY 14564 USA

chayes@bioworksinc.com

Keywords: IPM, integrated pest management, IPHM, integrated plant health management, *Trichoderma harzianum*

Summary

Nursery producers can use biologicals in integrated pest management (IPM) or Integrated Plant Health Management (IPHM) programs as stand-alone methods, or to complement chemical products for protecting plants from disease, insects, mites, nematodes, weeds – and other pests. Biologicals for plant pest control are derived from microorganisms, plant extracts, beneficial insects and organic matter. This paper describes how to properly use biologicals in IPM systems for green industry crops.

INTRODUCTION

Nursery producers can use biologicals in integrated pest management programs (IPM) as a stand-along method, or to complement chemical products for protecting plants from disease, insects, mites, nematodes, weeds – and other pests.
https://croplife.org/case-study/what-are-biological-and-why-are-they-important/

By definition, a plant pest is any species, strain, or biotype of plant, animal, or pathogenic agent injurious to plants. Examples include insects, mites, nematodes, fungi, viruses, bacteria, mycoplasmas, weeds, etc.

Agricultural biologicals are a diverse group of products derived from naturally occurring microorganisms, plant extracts, beneficial insects, and organic matter (Fig. 1). This paper reviews: why, which ones, when and where, how they work, compatibility, storage and shelf life, and tank mixing.

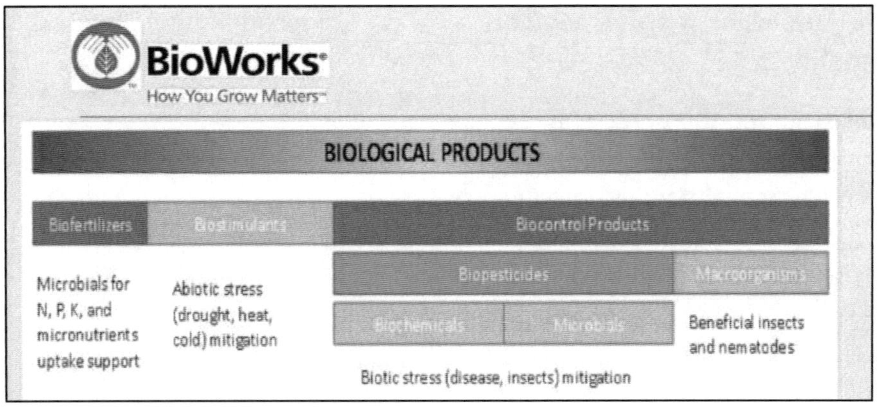

Figure 1. Categories of biological products used in IMP – integrated pest management systems of green industry crops.

Why Use Biologicals?

Biologicals as IPM and IPHM systems offer greater pest resistance than sole reliance on chemicals alone. They offer different modes of action (MOA), opportunities to change rotation of chemicals, greater safety of plants and personnel, faster reentry periods (REI of 0-4 hours) after application, viable use of beneficials (insects, microorganisms) – and it is environmentally friendly. Customers are looking for plant materials that are grown more sustainably. So, there are also enhanced marketing opportunities.

Biologicals are not a panacea. They are not effective under all conditions. They will not "cure" an outbreak/ high pressure population of insect pests or pathogens. They are more effective as preventative control and when pest populations are low and controllable. They are best applied early in a propagation or production program (Fig. 2). Always start with clean, uninfested plant material, and use cultural practices to keep a clean propagation and production environment.

Figure 2. Biological control, such as *Trichoderma* for controlling fungal pathogens is applied before pest problems occur. The use of uninfected plant material, and adaption of cultural practices to keep a clean propagation and production environment are equally important.

Appropriate Biological, Environment and Compatibility

It is important to use the appropriate biological for control of a specific pest or pathogen (Fig. 3). The propagation and production environment is an important consideration for effectiveness of the biological. Such variables as temperature extremes, pH, moisture and humidity all come into play. It is important to follow recommended usage and conditions for a specific biological.

BioWorks EPA Registered Biological Protection

Fungal	Bacterial
Control Root Pathogens *Trichoderma* spp.	Control Root Pathogens *Bacillus & Streptomyces* spp.
Control Foliar Pathogens *Ulocladium* spp.	Control Foliar Pathogens *Bacillus* spp.
Control Insects Above & Below Entomophagous Fungi - *Beauveria, Isaria, Metazhyzium*	Control Foliar Insects *Chromobacterium* spp.

Figure 3. Some examples of fungi and bacteria used as biologicals to control pests and pathogens.

Compatibility is also critical. How does the biological interact with other chemical and cultural practices in the propagation and production systems? What changes are needed to maximize the effectiveness of the biological?

Mechanisms of Biological Control

There are a number of mechanisms for biocontrol of pests. These include:

- Mycoparasitism (growth toward target fungi, lectin-mediated attachment, cell wall degrading enzymes) (Fig. 4).
- Predators and parasitoids – BCAS (BioControl Agents) (Fig.5).
- Production of secondary metabolites
- Competition for nutrients or space (Fig. 6).
- Tolerance to stress through enhanced root and plant development
- Induced resistance (Fig. 7).
- Solubilization and sequestration of inorganic nutrients.

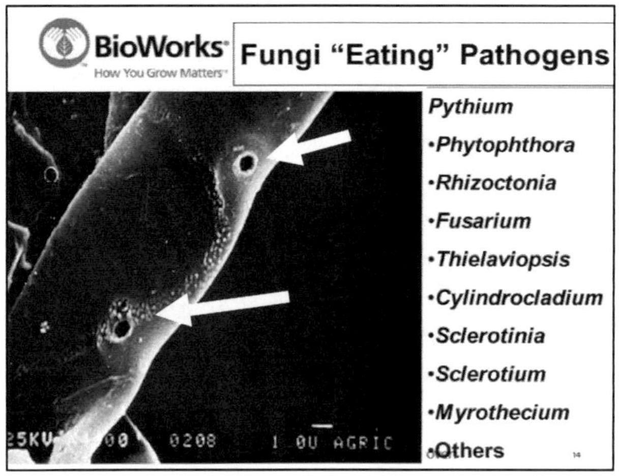

Figure 4. Biological solutions of beneficial fungi "eating" pathogenic fungi (arrows). https://www.ncbi.nlm.nih.gov/pmc/articles/PMC7734056/

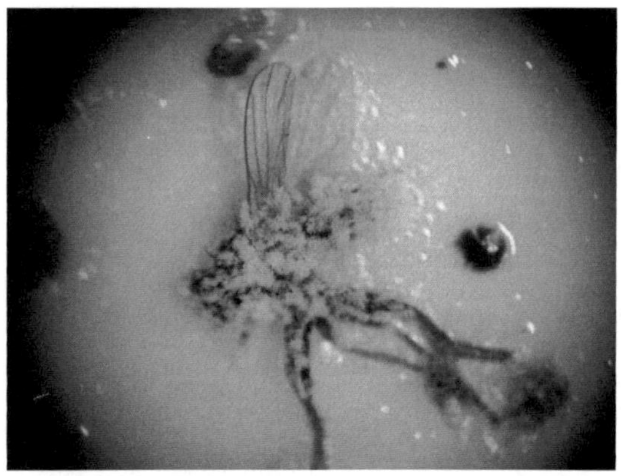

Figure 5. *Beauveria bassiana* strain GHA is feeding on a fungus gnat adult. Many strains of *Beauveria bassiana* fungi are found worldwide in the soil. They control insects by growing on them, secreting enzymes that weaken the insect's outer coat, and then getting inside the insect and continuing to grow, eventually killing the infected pest.

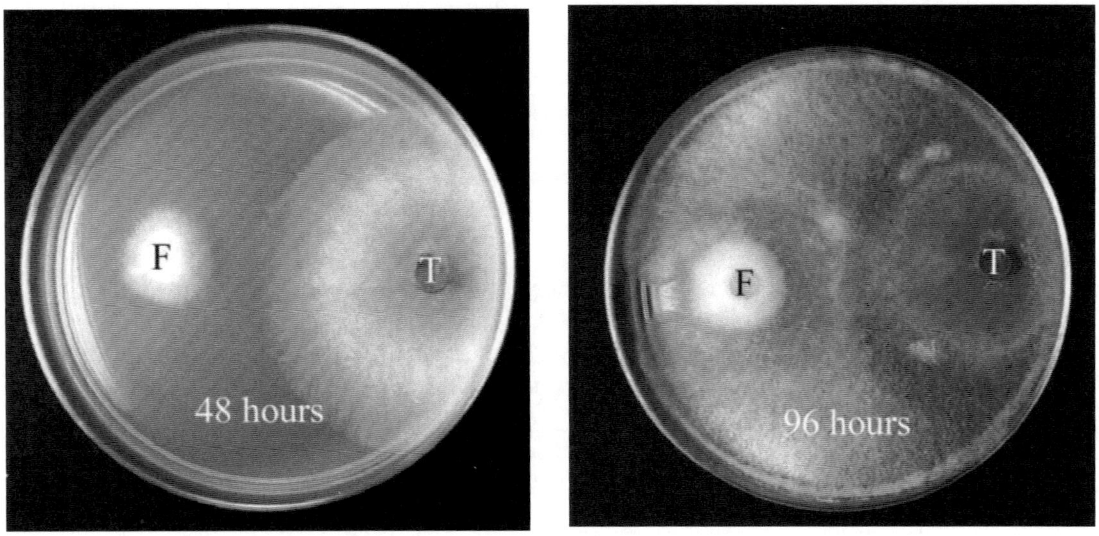

Figure 6. Speed is of the essence. An example of how rapidly the hyphae of *Trichoderma harzianum*, T22 (T) contains the pathogenic *Fusarium* (F) from 48-hours to 96-hours.

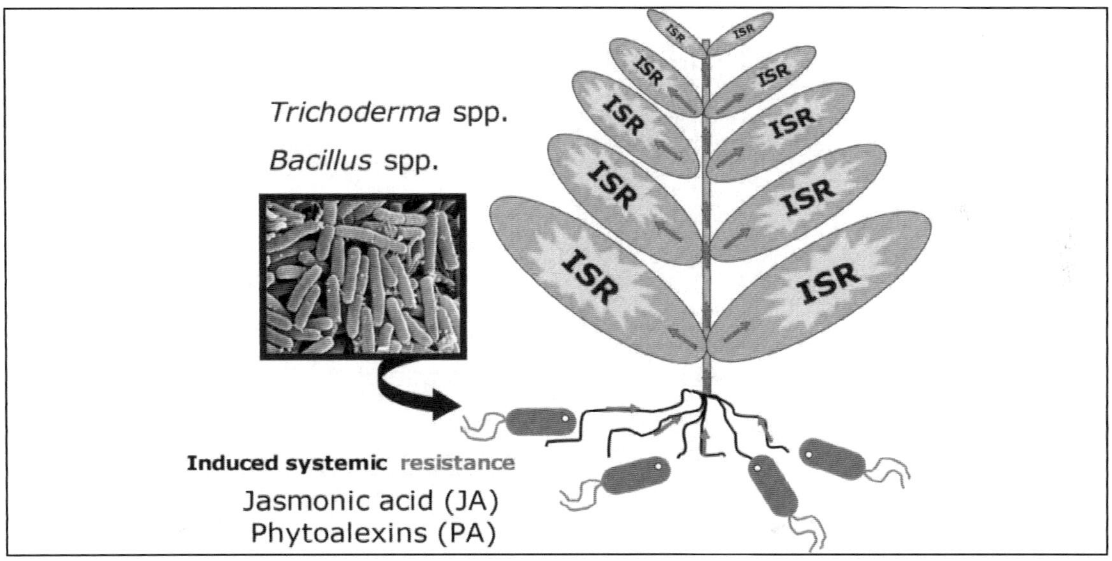

Figure 7. Biologicals such as *Trichoderman spp.* and *Bacillus spp.* can stimulate plants to have induced systemic resistance (ISR) to pests by producing phytochemicals such as jasmonic acid, and phytoalexins.

Do biologicals work?

Biologicals do work and can be preventative from weeks to months in controlling specific pests. The appropriate biological needs to selected to deal with the specific pest. The problem site is most effectively controlled when there is low to moderate pest pressure – not after a high-level outbreak of the insect or disease. It is important that the biological is compatible to your propagation and production system – thrives and grows. Effective coverage includes treating where the pest "hides".

Maintaining the proper conditions enhances the shelf-life longevity of the biological, i.e. avoiding high temperature conditions (Fig. 8).

BioWorks®

BioWorks® Technical Bulletin

Product Shelf Life

Insect Control

Product	Frozen (below 32°F)	Refrigerated (below 40°F)	Room (70-75°F)	Outdoors (above 75°F)
BotaniGard® 22 WP	Not needed[1]	Not Needed	12 mo[2]	Store below 85°F
BotaniGard® ES	Do not freeze[3]	Not Needed	18 mo[2]	Store below 85°F
Molt-X®	Not Needed[1] (keep at 20°F or above)	Not Needed	15 mo[2]	Store above 20°F, below 85°F
Mycotrol® O	Do not freeze[3]	Not Needed	12 mo[2]	Store below 85°F
NemaShield®	Do not freeze	About 4 weeks	Do not store at room temp.	Do not store outdoors
NemaShield® HB	Do not freeze	About 4 weeks	Do not store at room temp.	Do not store outdoors
SuffOil-X®	Do not freeze	Not Needed	2 yrs	Store cool

[1]Freezing will not harm product [2]From the time of manufacture [3]One time freeze/thaw will not harm product – check integrity of formulation if repeated freeze/thaw occurs.

Disease Control

Product	Frozen (below 32°F)	Refrigerated (below 40°F)	Room (70-75°F)	Outdoors (above 75°F)
CEASE®	Not Needed[3]	Not Needed	3 yrs	Store cool
MilStop®	Not Needed[1]	Not Needed	2+ yrs	Store cool
PlantShield® HC[2]	8 mo	6 mo	4 mo	1 mo or less
RootShield Granules with media	3 mo	5 mo	5 mo	2 mo or less
RootShield® Granules[2]	12 mo	9 mo	6 mo	2 mo or less
RootShield® Home & Garden[2]	8 mo	6 mo	4 mo	1 mo or less
RootShield® PLUS Granules[2]	14 mo	12 mo	6 mo	2 mo or less
RootShield® PLUS WP[2]	12 mo	10 mo	4 mo	1 mo or less
RootShield® WP[2]	8 mo	6 mo	4 mo	1 mo or less
T-22 HC[2]	8 mo	6 mo	4 mo	1 mo or less
RootShield AG	8 mo	6 mo	4 mo	1 mo or less
RootShield Seed Treatment	8 mo	6 mo	4 mo	1 mo or less
TurfShield® PLUS Granules[2]	14 mo	12 mo	6 mo	2 mo or less
TurfShield® PLUS WP[2]	12 mo	10 mo	4 mo	1 mo or less

[1]Freezing will not harm product [2]From the time product leaves BioWorks' warehouse
[3]Up to 3 freeze/thaw cycles will not harm product – check integrity of formulation if more than 3 freeze/thaw cycles occur (not tested beyond 3 cycles).

BioWorks • 100 Rawson Road, Suite 205 • Victor, New York 14564
800-877-9443 • www.bioworksinc.com

Shelf Life 4.30.2015

©2015 BioWorks, Inc.

Figure 8. Product shelf-life and storage of biological products.

IPM using non-living biologicals and cultural controls

There is also usage of non-living biological controls and cultural control agents. There must be good coverage for these products to be effective.

Examples of non-living biologicals and cultural control agents include:

- Salts
- Bicarbonates – Dehydrates & alters pH (Fig. 9)
- Oils & Soaps (Fig. 10)
- IGRs (Insect Growth Regulators – Residual Activity), i.e., Azadirachtin
- Pyrethrums - insecticidal compounds present in pyrethrum flowers.

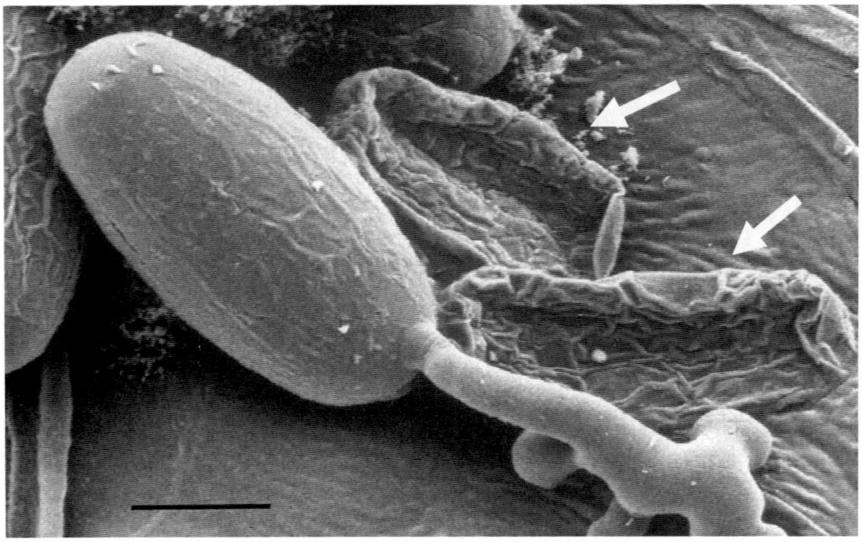

Figure 9. Bicarbonate treated (arrows) and untreated powdery mildew spores. The treated spores are desiccated and dead.

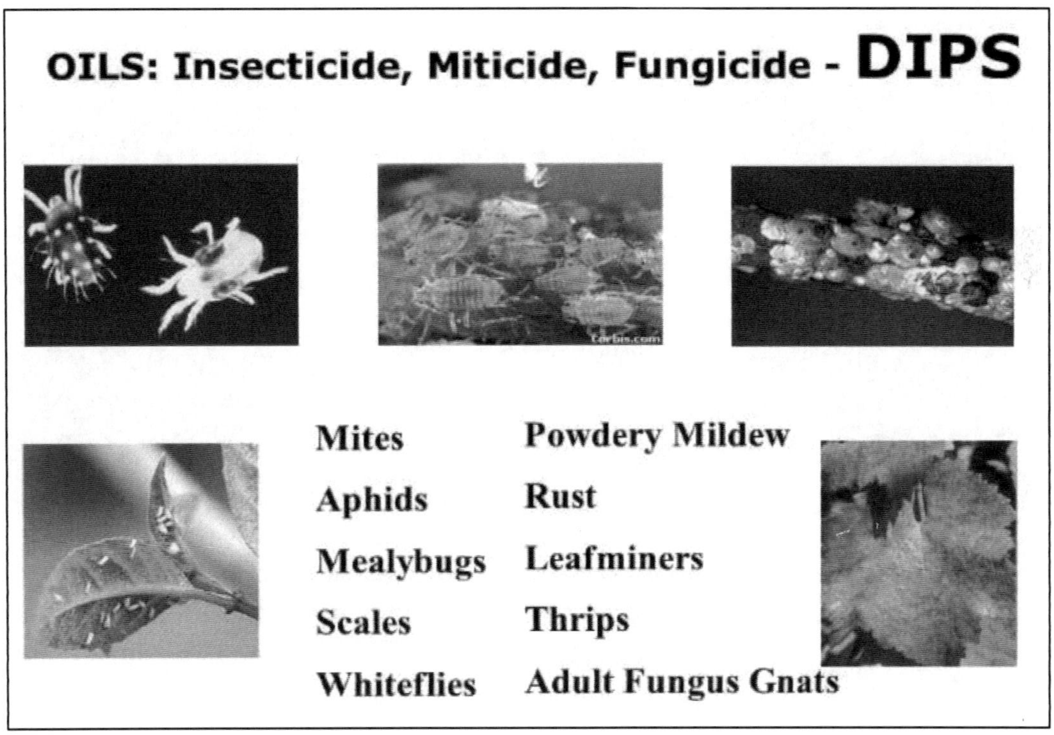

Figure 10. Various oils, petroleum and plant based, can be used as effective, safe alternatives to synthetic insecticides, miticides and fungicides. https://agrilifeextension.tamu.edu/library/farming/using-oils-as-pesticides/

Things to consider in tank mixing with biologicals and chemical products

Important issues to consider in tank mixing is the compatibility of the biological with other chemicals, and the need to adjust chemical mixtures. The compatibility between one biological and another biological is also important. And everything is about timing. This includes applying the biological at the appropriate time in the propagation and production schedule to control the pest. It also means the right season, i.e., horticultural oils are often best when the host crop is dormant. Also, releasing the biological during the appropriate time of the day and environmental conditions - so they can establish themselves and be effective control agents. As an example, parasitic nematodes released early enough during a controllable level of a pest population can be effective controls (Fig. 11).

Figure 11. Insect-parasitic nematodes (white curved structures) consuming an insect pest. Nematodes work in controlling pests such as with: *S. feltiae* - fungus gnat larvae, thrips pupae, flies; *S. carpocapsae* – armyworms, sod webworms, cut worms; *S. scapterisci* – mole crickets; and *H. bacteriophora* – white grubs, black vine weevils, borers.

CONCLUSION

Biologicals for plant pest control can be highly effective in IPM and IPHM systems when appropriately applied. Biology does work!

IPPS European Exchange 2019

Shea A. Keene

Department of Environmental Horticulture, University of Florida, Gainesville, Florida 32611-0670, USA

keene284@ufl.edu; shea.keene@gmail.com

Keywords: IPPS-Southern Region, Early-Career Professional International Exchange Program, English nurseries and gardens, Royal Horticultural Society (RHS) Garden at Wisley, IPPS-European Region Conference

Summary

In 2019, I was selected by the IPPS-Southern Region for the Early-Career Professional International Exchange Program to attend the European Region's annual meeting in Stratford-upon-Avon, England. The exchange program was one of the most amazing and impactful experiences of my entire life. I am so grateful for IPPS for giving me this once-in-a-lifetime opportunity, and for my wonderful hosts, IPPS International Chair Tim Lawrance-Owen and his wife, Annette. I have documented my exchange program experiences in this paper, including the gardens and nurseries I visited, the European Region's conference, and the final days of my trip in London.

INTRODUCTION

In 2016, I attended my very first IPPS Southern Region Annual Meeting in Virginia Beach as one of the Vivian Munday Young Horticulture Professional Scholarship recipients. Working the registration desk was an excellent way to meet many of the members and learn about the opportunities offered by IPPS, including the Charlie Parkerson Student Research Competition and the Early-Career Professional International Exchange Program. By the end of the meeting, I decided to become an IPPS-SR member. The following year, I presented a poster in the student research competition, and in 2018, I gave an oral presentation as one of the finalists in the student research competition and won first place. I applied for the exchange program both years but was not selected. Then in 2019, I finally achieved the goal I had been pursuing since that first meeting in Virginia Beach: I was chosen as the Southern Region representative for the International Exchange Program with the European Region. As they say, the third time's the charm! The exchange program was, without a doubt, one of the best, most amazing and impactful experiences of my entire life. It was truly a priceless, once-in-a-lifetime experience that I will treasure forever, and it was only possible because of IPPS. I am so grateful for the organization, and I am excited to document and share my experience as the 2019 SR representative in this paper.

In 2019, the European Region's annual meeting was held in England, in the charming and historic town of Stratford-upon-Avon. On the evening of October 1, 2019, I departed from the Tampa International Airport and flew through the night to the London Gatwick Airport. I hoped to sleep on the flight, but with my excitement level so high, sleep remained elusive. Just as I finally began to doze off, we encountered turbulence that lasted nearly an hour and put sleep firmly out of reach. Consequently, I arrived at Gatwick on Wednesday morning feeling quite exhausted and a bit disoriented. My host, IPPS International Chair Tim Lawrance-Owen, met me in the terminal and we walked to his car. To my mild embarrassment, I bumped into him as I automatically moved toward the right front seat. In my sleepless state, I forgot that was the driver's side! We laughed as I went to the correct side, then began the one-hour drive to Tim's home in Chichester.

As soon as we got on the road, the first thing I noticed was all the tall, thick, lush hedges. They were everywhere! I began to feel more awake as I marveled at the rolling green fields—bordered by tall hedges and crossed by low stone walls—and the architecture of the small towns through which we passed. We made several stops, the first of which was at a large garden center which vaguely reminded me of Rural King, but nicer and with more plants. I caught a glimpse of the English Channel from the top of a hill, and then we continued to the small town of Arundel where I saw Arundel Castle and Arundel Cathedral, as well as an incredibly old Catholic church that had been built in the 1300s. By that point though, the exhaustion was returning with a vengeance, so we continued onward and arrived in Chichester around mid-morning.

After a nearly four-hour nap, I felt much better. Tim and I had afternoon tea, then he showed me his lovely backyard garden and his incredible, envy-inducing Alitex greenhouse. His wonderful wife, Annette, returned home shortly thereafter, and we all went for a walk before dinner. We strolled down a lane to another lush, green, rolling field in which deer were grazing, then walked along a hedge-bordered road before cutting across a different field and returning home (Fig. 1).

We ate an early dinner, as Tim and Annette were having their church group over that evening. Their friends were all very sweet—one lady even said she loved my accent! But despite my long nap I was still exhausted, so I bid everyone a good-night and retired early. To my relief and delight, I had an excellent, full night of sleep and woke totally refreshed on Thursday morning, ready for my first nursery tour.

Figure 1. A lush, hedge-bordered field near Tim and Annette's home in Chichester.

Tristram Plants

We started my horticultural tour at Walberton Nursery, from which Tim had just recently retired after nearly 29 years of service. Walberton is one of three nurseries that make up Tristram Plants, which along with Toddington Nursery, form The Farplants Group, a cooperative of businesses and one of the largest wholesale suppliers of outdoor plants in the UK. Walberton mainly produces ornamentals and has a longstanding, active breeding program that has released many protected/patented varieties over the years. We first visited Walberton's propagation nursery, where we were given a tour by the propagation manager, Paul Dyer (Fig. 2).

Figure 2. Touring Walberton's propagation nursery with propagation manager Paul Dyer (top left). Outdoor production at Binsted (top right). Indoor production at Fleurie Nursery (bottom).

From there, we drove a short distance to the main Walberton Nursery, located in a huge complex alongside the other nurseries in the cooperative: Binsted Nursery and Fleurie Nursery. It was at the main nursery that I first noticed how few of the ornamentals I recognized. So many of their plants, like hellebores, just do not grow in Florida's hot and humid climate. And there were some plants, like hibiscus, that looked nothing like the tropical hibiscus to which I am accustomed in Florida! We joined some of the horticultural managers and staff of Walberton for their mid-morning coffee break, and then walked over to Binsted Nursery. Binsted produces herbs and bulbs, as well as alpines, succulents, and perennials.

After Binsted, we walked over to Fleurie Nursery. Fleurie had purchased that production site, which was formerly owned by Starplants, earlier in 2019. They were planning to expand and update the site, at which they primarily grew flowering plants. One of the last stops on the tour was the finishing and dispatch center, which serves all the nurseries in the complex. It was a vast warehouse space which, on dispatch days, would be bustling with plants and people; however, as we visited on a non-dispatch day, it was mostly empty.

RHS Garden Wisley

The next day, my horticultural journey continued at the Royal Horticultural Society (RHS) Garden Wisley (Fig. 3). Wisley is the flagship garden of the RHS, one of five gardens the RHS runs, and it is the second most visited garden in the UK after the Royal Botanic Gardens Kew. I must say though that after visiting both of those gardens, I found Wisley to be more impressive than Kew, and it was perhaps my favorite garden of the whole trip.

Shortly after arriving, we met up with Chris Moncrieff, the Head of Horticultural Relations at Wisley, and he gave us a little tour. He showed us the garden's major construction project: RHS Hilltop – The Home of Gardening Science, which is the UK's first dedicated horticultural science center. The construction was in full swing when we visited, and the center finally opened in June of this year (2021).

He showed us the new children's garden, which had been co-designed by The Duchess of Cambridge, and took us through the Glasshouse, including some of the "behind the scenes" areas. While in the tropical zone of the Glasshouse, I finally saw a lot of plants I recognized!

Figure 3. Visiting the RHS Wisley Garden.

Chris bought us coffee, and after chatting some more, he left us to continue exploring on our own. Tim and I spent the rest of the day meandering around the garden. They had an amazing bonsai collection, and I particularly enjoyed their Fruit Collections and Orchard, as well as the Rock Garden and Pinetum. We left Wisley late in the afternoon. On the way home, we stopped in the South Downs National Park—a range of chalk hills—and hiked a short distance to the top of a hill, which gave a great aerial view of a nearby town and lots of lush, hedge-bordered fields.

Historic Chichester

The following day, Saturday, Tim and Annette showed me around historic Chichester. I was stunned to learn that Chichester had been a Roman settlement as far back as 43 CE. We went to a neat museum that had been built on top of some Roman ruins. The middle of the museum was open and looked down onto the ruins, and all kinds of Roman artifacts and mosaics were displayed around the perimeter. Some of the original wall that surrounded the Roman settlement still survived, and you could walk along the raised walkway, which gave an excellent view of people's long, narrow backyard gardens.

We visited the Chichester Cathedral, which had existed on that site in one form or another since 1075. We ended the day with a walk around Wittering Beach. It was quite lovely, but as a native Floridian, I am a bit spoiled when it comes to beaches!

Vitacress

After spending all day Sunday working on homework and practicing my presentation for the conference, my horticultural journey resumed Monday with a tour of Vitacress. Vitacress is one of the largest producers of leafy greens and culinary herbs for UK supermarkets (Fig 4).

Figure 4. Cilantro (left) and basil (right) at Vitacress.

The production facility we visited was absolutely massive, covering multiple acres, and entirely indoors. It was at Vitacress that I first learned that the UK and European Union have much different and stricter regulations when it comes to pesticides, especially for edible crops. Rather than relying on conventional chemical controls, Vitacress had very strict cultural management practices and disease control programs, and they also used some biopesticides and other biological control agents. The general cleanliness and lack of pests/disease were especially impressive given the size of the operation. Vitacress also had very impressive equipment and automation throughout the facility.

West Dean Gardens and Drive to Stratford-upon-Avon

On Tuesday, I said my goodbyes to Annette, and we started the three-hour drive to Stratford-upon-Avon. On the way out of town, we picked up James, one of the 2019 "6-

packers," which is basically the European Region's equivalent of the Vivian Munday work scholarship program. Six young people (which is why they call them 6-packers!) are selected from a pool of applicants to work at the conference, helping set up, doing registration, introducing speakers, et cetera. With James in tow, we stopped at West Dean Gardens, the gardens on the estate of West Dean House, an old English manor (Figs. 5 and 6). While Wisley and Kew were obviously amazing, I think West Dean is tied with Wisley as my favorite garden of the whole trip. I absolutely loved it!

It was pure English, with beautiful rolling green hills dotted by grazing sheep, a sunken garden, a walled garden, Victorian glasshouses, a pergola, incredible espaliers, and lush hedges. It also had this great exhibit of old-timey gardening and lawn equipment, like old sprayers and an assortment of antique mowers and seeders. I wish we could have stayed longer, but with hours of driving still ahead of us, we had to get back on the road.

Figure 5. Part of the Walled Garden at West Dean Gardens.

Figure 6. Sheep grazing in the Parkland at West Dean Gardens (left). The 300-foot Edwardian Pergola at West Dean Gardens (right).

We arrived in Stratford in the late afternoon. I was introduced to Karl O'Neil, who was then serving as site chairman and Vice President of the European Region. I also met the rest of the 6-packers, and I helped them put together name tags and information packets for the conference attendees. That night, I had dinner with the conference committee members, 6-packers, and a few other early arriving members at a fantastic restaurant right on the River Avon.

IPPS European Region Conference

The conference itself was held the 9th thru the 11th, with presentations in the morning all three days, and nursery tours in the afternoon on the first two days. The 2019 theme was "technical times," and many of the presentations centered on the growing number of challenges faced by growers in the UK and EU. And just a note: the conference was held during a rather tumultuous time in the UK, right in the middle of the Brexit negotiations. The deadline to leave the EU had already been pushed back twice, and there were some pretty major terms of the withdrawal agreement that had not yet been finalized. It was very interesting to talk with the IPPS members about Brexit and hear the different opinions about it. I am curious how they feel now on the other side of Brexit, and if any of those opinions have changed!

The presentations covered a range of topics. One of the presenters spoke about developing biofungicides to replace some of the fungicides that could no longer be used, while another talked about helping growers get the best out of biopesticides to sustainably protect their crops. Another speaker covered a topic that is very relevant to US horticulture as well: trying to find a suitable replacement for peat in growing media and developing peat-free blends. A couple of presentations focused on using technology to improve efficiency and reduce inputs, especially optimizing water and fertilizer use. And there was an interesting presentation about the future of plastics in plant production, and the development of recyclable standard nursery pots in the UK. I gave my presentation right before lunch on the first day, which was perfect because I was able to introduce myself as the exchange program representative and share a bit about my academic and horticultural background with everyone at the same time. That put me in the unique position of being known to everyone at the conference, which made it much easier to meet the members. With perhaps 100 members in attendance, the conference was quite a bit smaller than the typical southern region meeting, so by the end of it, I think I spoke with everyone there at least once.

The first nursery we visited was New Leaf, a wholesale nursery that specialized in producing Clematis and climbing plants (Fig. 7). We were greeted by a very excited Labrador Retriever named Bella when we arrived, and she joined us for most of the tour.

From there, we went to Newey Avoncross, commercial contract growers of bedding plants (Fig.8).

Figure 7. New Leaf, specialist producers of Clematis and climbing plants.

Figure 8. A glasshouse of pansies at Newey Avoncross.

At that time of year, they were basically only growing violets and pansies. Their nursery was incredibly tidy and orderly, and they had some impressive equipment and automation as well. On the second day, we visited Bordon Hill, a production facility part of Ball Colegrave, which is Ball Horticultural Company's leading wholesale distributor in the UK (Fig.9). They grow ornamental plants from both seed and cuttings. The facilities were huge, and almost the entire nursery was connected by tracks so that benches could be moved from one area to another as needed. There was a very effective use of information systems and technology at the facility. Every tray or container was tagged with a barcode and tracked throughout its production life, and they had high-tech sensors throughout the greenhouses that monitored all kinds of environmental conditions, plus leaf temperature and vapor pressure deficit. Moreover, they had some of the most impressive equipment and automation I have seen in a nursery. They had a machine that filled trays with media and one that sowed seeds. Another scanned trays to detect cells in which seeds had not germinated, then it would remove the growing media from those cells with targeted puffs of air before filling them with a rooted plantlet. My favorite machine transferred plugs from trays to larger containers—it was mesmerizing to watch. The final nursery we toured was Hawkesmills Nurseries, which grows more than a thousand varieties of perennials, herbs, grasses, ferns, and vegetables (Fig. 10). They had just finishing constructing a brand-new glasshouse—it still had that shiny new feel to it, like a new car—and it was fun seeing so many different types of plants in one production space.

Figure 9. Poinsettias at Bordon Hill, part of Ball Colegrave.

Figure 10. The brand-new glasshouse at Hawkesmill Nurseries.

On the final day of the conference, there were a few more presentations in the morning, and then we all had one final lunch together in the early afternoon before everyone went their separate ways. It was bittersweet saying goodbye to everyone as lunch wrapped up, as I had really gotten to know some of the members in those brief few days.

Saying goodbye to Tim was the most difficult, as he had been such a wonderful host and guide during the first part of my trip (Fig. 11). But he had to get home to Annette, so we said our goodbyes and I was on my own.

Figure 11. Big smiles and bittersweet farewells with Tim on the final day of the conference.

London

After 10 days of meeting new people, networking, and constantly interacting with others, my introverted self was looking forward to spending the last few days of the trip on my own. It was also exciting because it was the first time in my life that I was traveling solo. I had decided to stay an extra night in Stratford to do a little sightseeing, but the first thing I did was look up a local laundromat, as I desperately needed to wash my clothes if I even hoped of getting everything back in my suitcase! Once that was done, I explored the town and did a bit of souvenir shopping. I saw Shakespeare's birthplace and a Shakespeare memorial with statues of his best-known characters, then strolled along the river before dinner.

I left Stratford early the next morning on a charter bus bound for London. Even though it took most of the day, I enjoyed the drive because I got to see parts of England I probably would not have seen otherwise. I arrived at my Airbnb in Hammersmith in the late afternoon, and after a full day of traveling, I decided to take it easy so I could get an early start the next day. On a whim though, I looked up the play *Harry Potter and the Cursed Child*. Amazingly, they had very reasonably priced tickets available for a showing the next day, so I bought them.

The next morning, I went to the Natural History Museum. It opened at 10 AM, which left me with only an hour and a half to explore before I had to leave for the play, so I picked one exhibit I really wanted to see: the Hall of Minerals and the jewel collection in The Vault. The two-part, five-hour play was at the Palace Theatre, a beautiful and historic theatre in the Westminster borough of London. The theatre was too far from the museum to walk, so I hailed one of the famous black London cabs, which made me feel rather fancy. When I found my seat, I could not believe how amazing it was: I was only a few rows back from the stage, and no one tall sat in front of me. The play was incredible—I cried at least twice!

On the penultimate day of my trip, I spent the morning at Kensington Palace, which was holding a special exhibit celebrating the 200th anniversary of Queen Victoria's birth. That afternoon, I made my way to the Royal Botanic Gardens Kew (Figs. 12 and 13). One of the 6-packers from the conference, Charles Shi, was currently interning at Kew, and had offered to give me a tour. I met him at one of the staff entrances, and he gave me a brief tour and bought me lunch, then left me to explore on my own. Of course, Kew was amazing! I especially enjoyed seeing the iconic Palm House, and the Temperate House was beautiful, too. I really enjoyed walking through the Princess of Wales Conservatory: it has 10 different climate zones, each containing a variety of plants that thrive in that climate. However, I was shocked to see a couple of plants growing—on purpose—at Kew. In the Princess of Wales Conservatory, they were growing cat's claw vine (*Dolichandra unguis-cati*). It became my archnemesis when I worked as a gardener at Bok Tower Gardens in central Florida, and it is a Category I invasive in Florida, which is the worst level of invasive. In the Waterlily House, they were growing what some people call balsam apple, but I have always called stink vine (*Momordica charantia*), and it is a Category II invasive in Florida. But I suppose one person's invasive weed is another person's botanic garden specimen!

Figures 12. The exterior (left) and interior (right) of the iconic Palm House at Kew.

Figures 13. The desert zone in the Princess of Wales Conservatory at Kew (top). Chihuly sculpture in the Waterlily House at Kew (bottom).

On the morning of my final day in England, October 15th, I went on a one-hour horseback ride through Hyde Park (Fig. 14). I am and always will be a "horse girl," so cantering through Hyde Park was absolutely magical.

I rode past the site of Winston Churchill's home, and I even saw several horses from the Queen's Household Cavalry being exercised in the park. Once my ride was over, I quickly packed the rest of my things and hustled to the London Victoria Station and took the Gatwick Express train to the Gatwick Airport. And just like that, my trip was over.

Closing Remarks

Words cannot adequately express how deeply and extraordinarily grateful I am that I was able to go on this trip. Every part of it was wonderful. I saw so many incredible places and things, met so many incredible people, learned so much about horticulture overseas, and made memories that I will cherish forever. It was truly a priceless, once-in-a-lifetime experience, an invaluable gift that IPPS gave to me. And then, when I think of how the whole world changed just a few short months later…I cannot even begin to describe how lucky, and thankful, and blessed, I feel to have had those experiences before COVID changed everything. I am so, so grateful for IPPS and its support of the exchange program, and I sincerely hope that as soon as things return to some semblance of normal, the Southern and European Regions will resume the exchange program, because it is truly a program worth supporting. Thank you, IPPS, for everything.

Figure 14. Horseback ride through Hyde Park on the final day of the exchange program.

Keeping Consumer Demand High: Riding the Green Wave of the COVID-19 Garden Explosion

Heather Kirk-Ballard

Louisiana State University AgCenter, 149 J.C. Miller Hall, LSU Campus, Baton Rouge, LA 70803, USA

hkirkballard@agcenter.lsu.edu

Keywords: Covid-19, home gardening, green industry, new gardeners, increased plant demand, increased sales

Summary

During the Covid-19 pandemic as the nation enforced a precautionary lockdown to help prevent the spread of infection, many households became restless. While other businesses were experiencing hardships because of shutdowns, the green industry was in high demand with many reporting significant gains in profits and sales. Garden stores reported high demand for vegetable crops, edibles, seeds and all gardening supplies and the Cooperative Extension Service also experienced an increase in demand for information on home gardening. In the 2018 National Garden Survey conducted by the Garden Media Group, 12 million Americans said, "I'm too busy to have much time for gardening." (Garden Media Group, 2018). Now with time on their hands, Americans began gardening in droves and this is good news for the green industry.

INTRODUCTION

What do you do with time on your hands? Survey says: gardening. When the Covid-19 pandemic forced Louisiana and the rest of the nation into a quarantine scenario in the spring of 2020, many people found themselves with extra time on their hands. During this time, the United States gained 16 million new gardeners with many of them under the age of 35. Adults were spending two additional hours a day outside during quarantine than before with 84% of Americans spending more time in their gardens than before the pandemic (Garden Media Group, 2021). Retailers reported a 10% increase in plant purchases from the year prior and additional research in 2020 revealed a 4% increase in consumer spending on plants compared to 2018 levels (Grassi, 2021).

With gyms, entertainment venues and restaurants closed and people no longer having to commute to work, a surplus of time remained. In addition, supply chains were uncertain, food availability was a concern, and people did not want to spend one more minute indoors, so they turned to gardening in record numbers. There was anecdotal evidence all across the state and the rest of the nation that indeed people were gardening. Nevertheless, was this real? I personally witnessed it. Neighbors who in the past paid a company to maintain the lawn were now out in their lawns tackling garden projects or putting in a vegetable garden for the first time. People just wanted to get out of the house and use this newfound time to be productive, physically active and get some fresh air.

As the consumer horticulture extension specialist for Louisiana, I wanted to understand how the forced quarantines affected the amount of gardening consumers did. A consumer gardening survey was put together asking participants all across Louisiana about their gardening habits during the pandemic, the buying activities of those consumers from their own perspectives, and the perspective of the retail garden centers.

MATERIALS AND METHODS

A survey entitled Gardening in Louisiana during Covid-19 was developed containing 17 questions and was distributed to people across the state of Louisiana. The Institutional Review Board (IRB) Determination of Exempt Human Subjects Research was applied for and approved through Louisiana State University (#HE20-39). The survey was delivered using Qualtrics and distributed through email listings from extension offices and LSU AgCenter social media outlets. Participants were 18 years and older and of every demographic (N=2,195).

RESULTS

The survey asked participants how long they have been gardening, how much time they spent in the garden prior to and during the stay-at-home orders, and how much time they expect to spend in the garden after the COVID-19 pandemic.

The survey revealed that 59% of participants were already avid gardeners and had been gardening for more than 10 years, 11% for five to 10 years, 17% for one to five years, and 10% said they were first-time gardeners. The most exciting find for me was that 10% were first-time gardeners. Next, participants were asked if they had increased the amount of time they spent gardening during the pandemic. Results showed that 82% of the participants said they had increased the amount of time spent gardening.

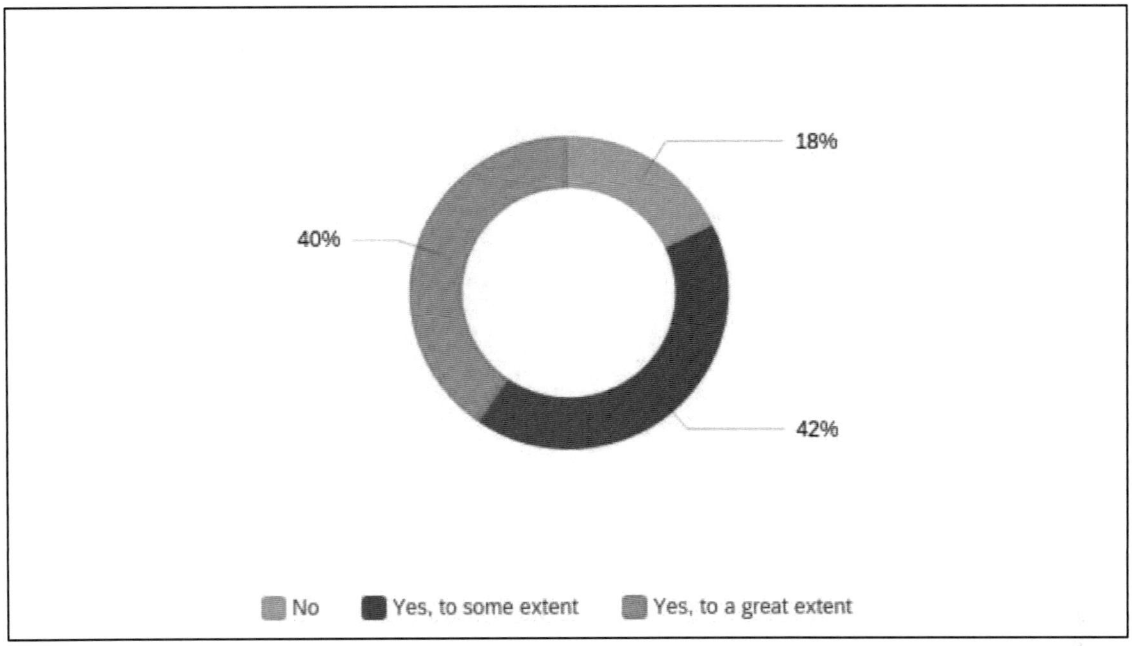

Figure 1. Percentage of gardeners who increased the amount of time they spent gardening. A total of 82% of gardeners said they increased the amount of time spent gardening during the COVID-19 pandemic.

When asked how often they were gardening during the pandemic, 46% said they were gardening five to seven days a week and 35% said they were gardening three to five days a week. One of the most important lessons learned from the survey might be the reasons people gave for gardening. The No. 1 answer was "because it makes me happy" at 88%. Other answers included stress relief and relaxation at 88% and to be out in nature and outdoor physical activity at 87%. This confirmed my belief that, indeed, most people garden because of its well-being benefits.

Many studies have shown that gardening activities have been associated with enhanced positive emotions and decreased negative ones (Theodorou, 2021).

Additionally, the survey indicated that 81% of participants plan to continue to garden at this rate after the pandemic. The results are promising for the green industry, and that is good news for everyone involved. These results align with another recent gardening survey conducted by Axiom Marketing that found 86% of homeowners plan to continue gardening in 2021.

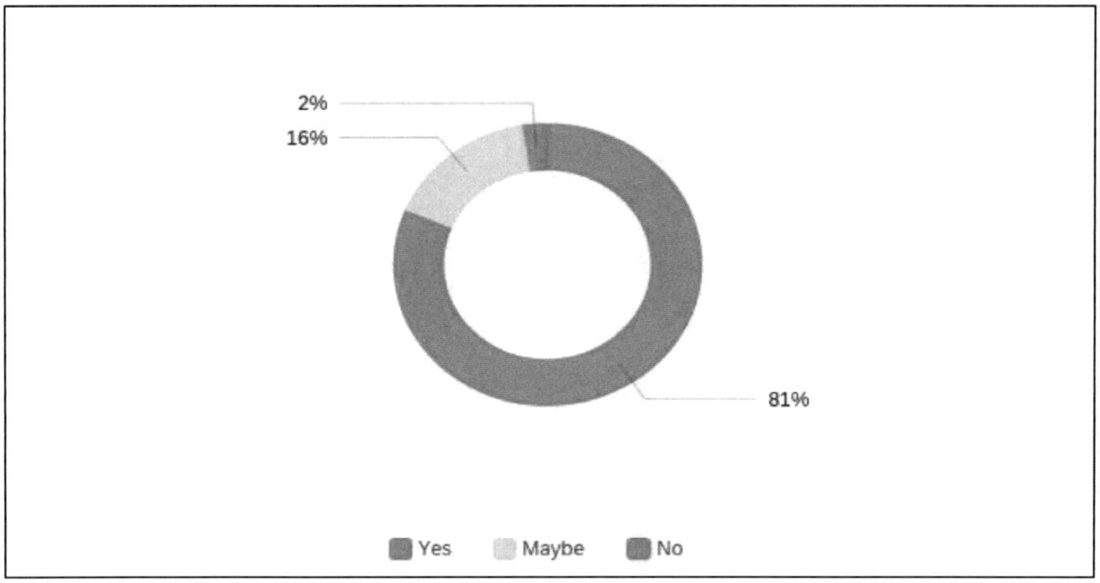

Figure 2: Percentage of gardeners that plan to continue to garden and the percentage of those who will not. Survey results indicated 81% of gardeners said they would continue to garden when things went back to normal after the pandemic.

Finally, the survey asked participants how they found the information they needed on gardening during the pandemic, and 51% of Louisianans said they found their information from the LSU AgCenter website, social media pages, news articles and local agents — another great find from this research.

Throughout history, the Cooperative Extension Service, which is administered through land-grant universities in this country, has answered the call to help people, and 2020 was no exception. We have seen throughout history in events such as World War I, the Great Depression and World War II when Cooperative Extension has worked to support rural and urban Americans learn to grow and preserve food.

DISCUSSION

Several trends have emerged from the pandemic. Perhaps one of the most profound has been that young homeowners have led to a gardening boom in 2020. According to a survey conducted by Axiom Market entitled "Axiom 2021 Gardening Insights Survey: Gardening in a COVID-19 World" ages 19-28 and 29-39, 57% said they visited garden centers more during the pandemic and spent the highest mean on plants than the rest of all age groups. Of the age group 19-28, 90% of participants said they felt successful with gardening. So, how do we keep the momentum going? By keeping these gardeners successful. There is a strong trend for local buying. E-commerce is up and sustainability is a market driver. Customers want to see products that are sustainably produced. That means reducing your carbon footprint, reducing the use of plastics, plant tags and the increased use of sustainable production practices such as reduced chemical use and water conservation with the planet in mind.

CONCLUSIONS

The new trends are here to stay. As an industry, we need to take what we learned during the pandemic and run with it. The supply chain has changed forever from all in-person shopping to delivery and curbside pickup. Successful businesses will shift from the old models and be flexible to quick change. Consumers want convenience. The NGA survey of 2019 indicated that 10.4% of all garden and grocery, shopping was done online and curbside pick-up increased to 90% during the pandemic. One-quarter of Americans spent more money while social distancing from the comfort of their homes and that is a trend that is likely to stick. Bottom line, customers want convenience and speed with an overall goal of efficiency.

We are in an economic boom, the National Retail Federation (NFR) anticipates that retail sales will grow at least 10.5% in 2022, surpassing initial estimates at $4.44 trillion and it will be the strongest year since 1984. If you want to get on the bandwagon, you have to get with the times. Social commerce is expected to reach 4.3% of retail e-commerce sales in 2022 at $36.09 billion.

What do those consumers want from the green industry? Container plants that are attractive and that are low maintenance. The container gardening trend is hot right now. People have smaller spaces to work with and they need options for growing. Containers can be the answer for many consumers. There will be a sustained growth in house plant demand and an increase in edible plants is expected in the years to come. Consumers also are gardening with wildlife in mind. In 2020, 67.2 million households purchased at least one plants because it benefited pollinators or birds.

Lastly, gardeners want everything at their fingertips. They want quick information they can find on tablets and phones. The demand is high and the momentum is up. Let's ride the green wave.

LITERATURE CITED

Axiom 2021 Gardening Insights Survey: Gardening in a COVID-19 World: 2021 Garden Survey. https://bit.ly/Axiom2021GardeningSurvey

Garden Media Group 2018 Garden trends report 2019: Rooted together. 17 Dec. 2019. http://grow.gardenmedia-group.com/2019-garden-trends-report.

Garden Media Group 2021 Garden trends report 2021: The Great Reset. 14 Sept. 2020.http://grow.gardenmedia-group.com/2021-garden-trends-report.

Grassi, Matthew J. Tactical economics. Nursery Management. 8 Feb. 2021. https://www.nurserymag.com/article/green-industry-economic-predictions-for-growers/

Theodorou, A., Panno, A., Carrus, G., Carbone, G.A., Massullo, C., and Imperatori, C. (2021). Stay home, stay safe, stay green: The role of gardening activities on mental health during the Covid-19 home confinement. Urban Forestry & Urban Greening 61:1-6. https://doi.org/10.1016/j.ufug.2021.127091

Is Your Fertility Program Stuck in the 1950's? Well, It Should Be! Nutritional Approach to Passive Plant Pest Immunity

Michael Roe
Windmill Nursery of Louisiana LLC, 12398 LA-25, Franklinton, LA 70438 USA

Michael@windmillnurseryllc.com

Keywords: Brix, disease and pest resistance, fertilizer systems, integrated pest management, microbiology, mineral nutrition, plant health, point of deliquescence, reducing plant stress

Summary

Many green industry production systems are not sustainable and inherently wrong. Changes are needed in our nutritional and chemical programs that enhance sustainability: environmentally, economically and culturally. It is possible to manage plant nutrition in such a way that plants become more resistant to insects and diseases. The plant health pyramid from Advancing Eco Agriculture® illustrates what we are trying to achieve in terms of plant growth and health. Mineral nutrition and microbiology are the foundation of plant immunity and pest resistance.

Base Saturation or "ideal ratios of cations" in the soil/ container media are critical for balanced plant nutrition, health and pest resistance. Testing the leaf Brix index of plant sap with a refractometer is a quick way to determine plant health. The leaf Brix level/insect relation chart is an excellent tool for gauging plant health and pest resistance. A low leaf Brix level (0-6) indicates plant susceptibility, whereas plants are largely resistant to insects and disease at Brix levels 12 to 14. Optimizing the plant nutrient levels and minimizing pesticide usage - can significantly increase photosynthesis, Brix levels – and increase pest and disease resistance.

INTRODUCTION

In the 2006 movie, "Idiocracy", the lead character wakes up one day to 500 years in the future. He quickly realizes that he is the smartest person on the planet - as people have been completely dumbed down by inferior education, misinformation, propaganda and advertising. Farmers are irrigating their fields with a sports drink called Brawndo because it contains what plants crave: "Electrolytes". The lead character quickly gets them to start irrigating with water and both the crops and civilization are saved. Sounds like the most ridiculous thing you've ever heard right? No way that could ever happen in a civilized society? Well, something very similar has been happening in agriculture for decades – for over a century!

What we are going to do right now is screen a 60-sec move trailer for the sequel to this movie. Our lead character/ hero has now woken up in present day 2021 and is sitting in the audience with you. We are simply going to show him a few slides depicting how we currently grow plants. The question is how long do you think it will take – for our hero to realize that once again, he is the smartest person on the planet?

ENVIRONMENTALLY UNSUSTAINABLE PRODUCTION SYSTEMS

Cotton is considered one of the 'dirtiest' crops on the planet in terms of the amount of chemicals and pesticides used to grow it (Fig. 1). What if we tell our hero that we make clothes, bed sheets and bath towels out of this cotton? Then the rest of the plant is utilized for cooking oil and animal feed. https://www.moderndane.com/blogs/the-modern-dane-blog/why-cotton-is-called-the-worlds-dirtiest-crop

Figure 1. Cotton Farming is universally recognized as one of the most chemical-laden crops on the planet.

What if we showed him how many chemicals it takes us to produce unblemished fruit and vegetables that we consume and feed our children? What if he saw that we were destroying our pollinators by spraying all these chemicals? In some areas of China, they have killed off most of their pollinators. They literally take a sack of pollen and a stick with chicken feathers on it and climb up into the fruit trees and pollinate each of the flowers (Fig. 2). I was born and raised in central Florida so citrus is near and dear to my heart. What if he saw the methods we have resorted to keep insects off of our citrus trees (Fig. 3)? Is this really the best we can come up with as craftsmen of our trade? Is this the best we can come up with as a civilization? I think not!

Figure 2. Hand pollination of temperate fruit crops is done in parts of China because insect pollinators have been killed off. Credit: Huffington Post.

Figure 3. (right) Citrus under protective screen (CUPS). (left) Steel roll-up door with a second plastic strip door, and (right) citrus can be grown under protective screen structures for fresh-fruit production in order to exclude the Asian citrus psyllid (ACP, *Diaphorina citri*) vector of huanglongbing (HLB), or citrus greening disease, and thereby produce disease-free healthy fruit. While CUPS may be economically feasible for some fresh-market citrus crops, over 90% of Florida citrus is for juice production, so a CUPS system is not economically sustainable. https://edis.ifas.ufl.edu/publication/HS1304 Credit: Arnold W. Schumann, UF/IFAS.

Many production systems are not sustainable and inherently wrong. It defies common sense and logic. Yet we all do this every single day and think nothing of it. This is all we have known because it is all we have seen and been taught by others. It is what our educational system teaches us. It is what the fertilizer and chemical companies' market. I grew up gardening with my father and grandfather and literally thought the pesticide Sevin dust was a tomato fertilizer! It does not have to remain this way because long-term it is not sustainable: environmentally, economically or culturally.

HORTICULTURAL REALITY

What comes next is what I refer to as 'Horticultural reality' and unfortunately it is probably the first time you have ever heard some of this information. This will be the shortest, least expensive and most rewarding "PhD course" of all time. Our nursery production company is not immune to common commercial problems. We still get aphids, chili thrips, whiteflies and flea beetles. However, what were once ginormous problems have been reduced to minor inconveniences because of sustainability changes we have made in our nutritional and chemical programs.

There is a better way. It is possible to manage plant nutrition in such a way that plants become more resistant to insects and diseases. This is not a theory or hypothesis – it is something that has been executed on millions of acres. When you manage nutrition sustainably the entire system begins to behave differently. Most of the information from this point forward was gathered from books written in the 1940's through 1960…hence the title of this paper.

The plant health pyramid from Advancing Eco Agriculture https://www.advancingecoag.com illustrates what we are trying to achieve in terms of plant growth and health (Fig. 4).

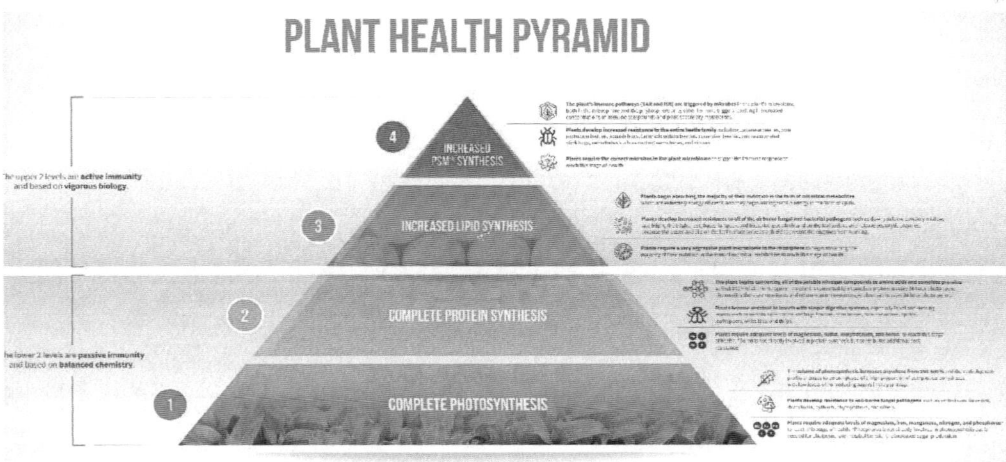

Figure 4. The Plant Health Pyramid. The lower two levels are passive immunity and based on balanced chemistry. The upper two levels are active immunity and based on the vigorous biology of a healthy plant. Mineral nutrition and microbiology are the foundation of plant immunity and pest resistance. AEA – Advancing Eco Agriculture https://www.advancingecoag.com

There are the top four levels of plant health and infinite levels below them. This paper focuses on the bottom two levels of the pyramid, which are entirely dependent upon a "balanced chemistry" and nutritional program.

SYSTEMS APPROACH

We're going to take a 'Systems Approach' to plant production. Our system includes soil, weeds, insects, diseases and nutrients – and their influence on nursery crop production. Optimum performance levels of the soil (media) are required to grow a healthy crop. It begins with base saturation which means proper cation saturation or the ideal ratios of cations. The base saturation of soil may be the most important figure in this paper (Fig. 5).

Base saturation of soil

- ❑Developed by William Albrecht at University of Missouri from 1930's – 1950's.

<u>'Ideal Soil' Cation ranges below:</u>

- ❑Calcium 68% (60-70%)
- ❑Magnesium 12% (10-20%)
- ❑Potassium 5% (2-7%)
- ❑Sodium 1% (anything less than 2%)
- ❑Hydrogen 10%
- ❑All other Cations combined (minor elements copper, iron, manganese, etc) 4%

Figure. 5. Dr. W. A. Albrecht saw a direct link between soil quality, food quality and human health. He drew connections between poor quality forage crops, and ill health in livestock. From this he developed a formula for ideal ratios of cations in the soil - the "Base Cation Saturation Ratio". Albrecht was also one of the early proponents of developing agroecology for more sustainable production systems.

The base saturation levels for optimum soil conditions and plant growth are 68% calcium, 12% magnesium, 5% potassium, 1% sodium, 10% hydrogen, and 4% all other cations. The most important concept to understand is that the soil is a living system with plants and rhizosphere organisms; what we are trying to accomplish with base saturation is: 1) creating a hospitable environment for beneficial microorganisms to thrive in and 2) creating an environment where all of the mineral elements are in ratios that optimize the nutritional needs of producing a crop.

If you balance the cations - you can throw your pH meter out the window! Your soil should always be around 6.4 pH when cations are properly balanced. A 6.4 pH is also the ideal pH of sap in a healthy plant and the pH of saliva and urine in a healthy human. If the mineral balance is in-line, the pH will self-correct. [Controlling the pH of irrigation water is also critical for sustainable nutritional programs].

Once the soil minerals are balanced and part of a living soil, the reliance on insecticides and other toxic "rescue" chemicals is reduced. Living organisms that are nutritionally fit have greater resistance to disease, parasites and other pests. While pH means acidity to some people, it is also an indicator of shortage of fertility elements; the same fertility elements responsible for base saturation.

CALCIUM (CA): MAGNESIUM (MG) RATIOS

Calcitic limestone has a 35:1 Ca:Mg ratio, while dolomitic limestone has a 2.5:1 Ca:Mg ratio (Fig. 6). You cannot get to a 7:1 Ca:Mg ratio by using either alone. In fact, if you were to solely use dolomite you would reach pH 6.4 long before you ever added the ideal amount of calcium, since magnesium is 40% more effective at raising pH than calcium is. With just calcitic limestone you could never get sufficient magnesium into the soil. So, a combination of both, or either with some additional amendments is required to reach the proper Ca:Mg ratio.

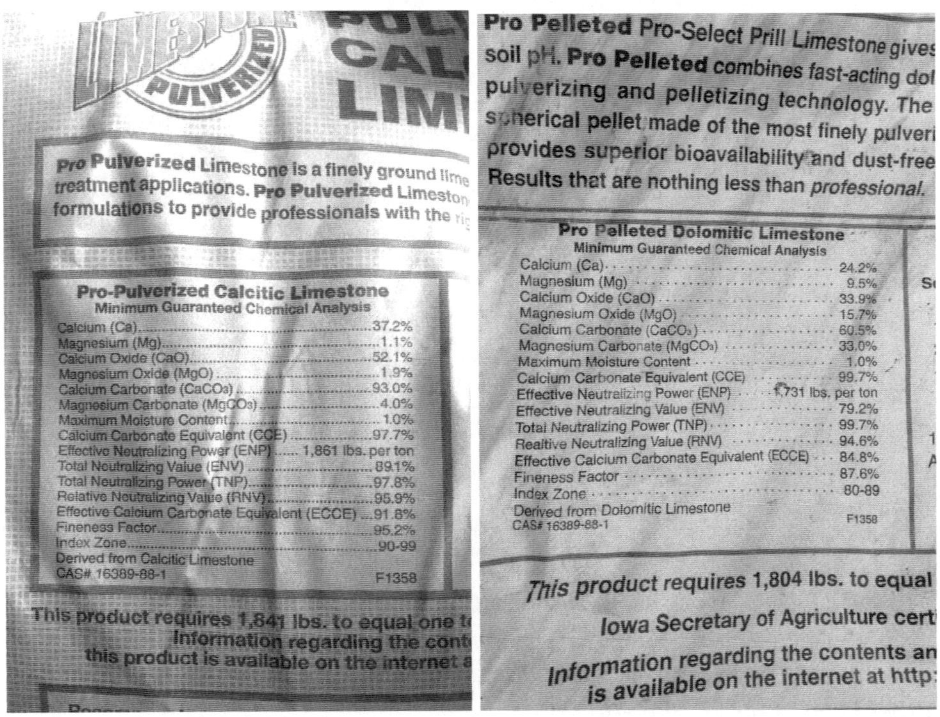

Figure 6. Nutritional components of calcitic limestone (left) and dolomitic limestone (right).

A balanced equilibrium of calcium and magnesium creates a sustainable soil environment for bacterial and fungal activity for proper decay of organic residues into CO_2, carbonic acid and a host of many weak and mild organic acids - all critical to convert and release mineral elements in the soil system. The Ca:Mg ratio sets the stage for the rest of the elements. If calcium is too high, the soil tilth will be loose, but will lose its texture and cohesiveness and water may percolate through too quickly. If magnesium is too high, the soil will be tight, restricting water and air movement.

At no time do we want the calcium to drop below 60% base saturation, or magnesium to drop below 10% - unless we are growing specialty crops such as blueberries or rhododendrons that like a high Mg and acidic soil, or certain plants that prefer a very high calcium soil.

Calcium. By weight and volume, calcium is needed more than any other element. Calcium is essential for its energy creation potential in the soil to release the other elements that cause a plant to grow. It's basically like a nuclear power plant within the soil. Calcium has the leadership role among the other essential plant nutrient ions. As protein concentration rise, calcium concentration also rises. And with an increase in protein there is also an increase in vitamins. When a small amount of calcium in 100% soluble form is introduced to the soil, its energy power is far greater than any of the other elements surrounding it. If there is calcium in the soil and some moisture, the new input has the tendency to convert insoluble calcium in the soil and into a soluble form.

Magnesium. Magnesium controls the total amount of nitrogen in the leaf so that an excess does not build up. Magnesium is the main governor of nitrogen in the soil (as such, it is an excellent antidote for nitrogen toxicity). Magnesium, pound for pound, can raise pH up to 1.4 times higher than calcium. Magnesium has a favorable effect on the movement of nitrogen and phosphorous into plants. Excessive magnesium will cause phosphorous, potassium and nitrogen deficiency.

The ratio of phosphate to potassium (potash) in the soil should be 2:1 (1:1 elemental P:K) on crops other than grasses. In grasses, this ratio should be 4 phosphate:1 potassium. If potassium is replacing calcium in the leaf, both the leaf and the stem will exhibit small black specks. When excess potassium is applied, it replaces calcium and plants are more susceptible to disease. Potassium seems to have a more pronounced effect on disease caused by soilborne fungi than on diseases caused by airborne fungi.

WEEDS CAN BE AN INDICATOR OF A SOIL'S HEALTH

Weeds can be an index of what is wrong with the soil, or at least with the fertility program. https://www.canr.msu.edu/news/weeds_are_an_indicator_of_a_soils_health Field horsetail is a good indicator of poorly drained, low pH soils. Improving the drainage and increasing the soil pH by liming will help to manage field horsetail as a weed. Weeds may be indicator plants for soils low or high in certain elements. For example, Quackgrass (*Elytrigia repens*) is a sign of improper iron-manganese ratio.

Broadleaf weed pressures can often be controlled by balancing phosphorous and potassium levels: 2 phosphate - 1 potassium are ideal. As available nutrient ratios drift from (2 Phosphate: 1Potash) broadleaf weed pressure will increase. When ratios get to 1:8 and beyond, the weed pressure becomes so intense that herbicides become ineffective.
Some examples of conditions preferred by some common weeds are as follows: prostrate spurge and yellow nutsedge both prefer low levels of calcium and phosphate, and high levels of potassium and magnesium. Crabgrass, ironweed, morning glory and dandelion are indicator plants for soil deficient in calcium. Most old timers know the best herbicide for dandelions in the yard is an application of calcitic limestone.

Grassy weeds generally indicate calcium deficiency. They are also present if the magnesium is too high in relation to calcium. It's a pretty safe assumption that if you have grassy weeds, additional calcium is needed in your soil.

Optimum fertility can mitigate weeds. See the example of a healthy young Buddleia and a weed in the same container - succumbing to a foliar pathogen (Fig. 7). The weed being attacked by a pathogen is an indicator that the container nutrition program is balanced. The intended crop is healthy and pathogen-free, while the weed is susceptible to the fungal pathogen. [A dilemma for the green industry is zero tolerance for weeds in container production systems. Weeds in container production are a costly management problem to control via integrated pest management, chemical usage, and expensive hand-weeding].

Figure 7. (left) A healthy young *Buddleia* and (right) a weed (arrow) growing in the same container as the *Buddleia* but gradually succumbing to a foliar pathogen.

NUTRITIONALLY DEFICIENT AS WELL AS UNBALANCED, "OVER-FED" CROPS ARE MUCH MORE SUSCEPTIBLE TO INSECT PESTS AND DISEASE

According to William Albrecht: "Insects are nature's garbage collectors and diseases are her cleanup crew". It may be a new way of thinking about things, but I assure you this is 'horticultural reality'. With a malnourished, unhealthy plant - insects and diseases can proliferate – ultimately 'taking out the trash' with the plant's demise. So, if you see insects and disease on your plants, they are indicators of a stressed, susceptible crop.

Phill Callahan was an Entomology professor at the University of Florida and also worked for the USDA. He published a dozen books on insects and retired from the USDA a mere decade before I went to school there. A lot of his work was on insect communication and how they see in the infrared spectrum. So, all the chemicals we spray on plants combined with unbalanced nutrition creates an environment where our plants outgas ethanol and ammonia; insects recognize this in the IR spectrum as well as sense with their antenna - and are drawn to stressed crops as if it were an "all-you-can-eat buffet".

TESTING THE LEAF BRIX OF PLANT SAP WITH A REFRACTOMETER IS A QUICK WAY TO DETERMINE PLANT HEALTH

The Brix index is measurement of carbohydrates and can be measured with a digital handheld refractometer (Fig. 8). Degrees Brix is the carbohydrate/ sugar content of an aqueous solution. Leaves can be crushed via any method (i.e. garlic press), leaf sap expressed, placed on top of the lens, and a reading taken. It is very simple. It is a very basic concept with very basic requirements. Plants require sunshine, water, air in the soil, and nutrients to photosynthesize and create high levels of carbohydrates.

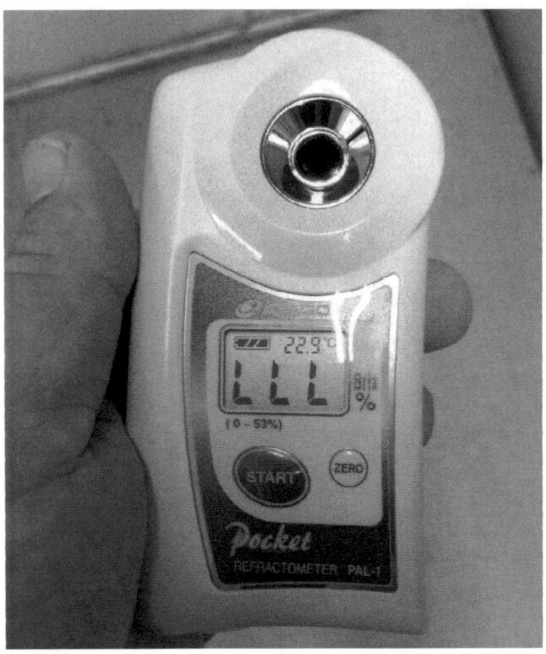

Figure 8. A refractometer is used to measure the leaf Brix which is the carbohydrate/sugar concentration as a percentage. Testing the Brix leaf sap with a refractometer is a quick way to determine plant health.

What contributes to lowering Brix is not always simplistic. All the "-cides": insecticides, fungicides, herbicides can lower plant Brix. A nasty side effect is that different insecticides, fungicides and herbicides can kill very specific groups of microbes as well. So, like most of us in the green industry – there is massive rotation of insect, disease and weed sprays. In essence you are doing your best to annihilate most of the beneficial microbes inhabiting the soil as well as those colonizing the leaf, shoot and root surfaces. Sounds like the exact opposite of what we should be doing - doesn't it? High salt fertilizer will also lower Brix.

What does Brix have to do with insects? The leaf Brix level/insect relation chart was developed by Dr. Thomas Dykstra of Dykstra Laboratories, Inc. in Gainesville, FL. https://www.ecofarmingdaily.com/grow-crops/picky-eater-insects-pass-on-high-brix-plants/

His lab opened in 1997, two years after I arrived at UF. According to Dykstra: insects have a simple digestive system and cannot digest the same foods that we do. Low-Brix plants (6 and below) are easier for the insect gut to digest. Insects lack the digestive enzymes to break down healthy proteins from high-Brix plants. Essentially a high Brix (14 and above) means that insects will not attack a given plant, nor are they attracted to it.

If your plants have aphids, scale, or mealybugs they are in essence on "nutritional life support". If left alone these plants would succumb to insects and disease and be "taken down to the ground". If one raises the Brix levels above 6, insect pest species change, i.e. thrips and whiteflies, etc. dominate; at higher Brix the next level of herbivore species include caterpillars, worms, and beetles. [Once most plants reach 6 Brix, there is a significant jump in the production of secondary plant metabolites, which are phytochemicals that help contribute to a plant's odor, color, and provide natural plant defenses against pests. These 6-Brix plants are finally able to devote their energy reserves into producing new proteins and more diverse molecules]. The final pests are the grasshopper group. Chewing insects that eat the roots or leaves directly, such as caterpillars, grasshoppers, and beetles, will start to lose interest in eating a plant with a 10 or 11 Brix. At Brix level 12 to 14 - plants are largely resistant to insects and disease.

Up until a few months ago I had no idea that there were different levels of plant health that correlated with different insects. So, for years after foliar spraying nutrients to get rid of aphids and then had whiteflies or caterpillars arrive, I thought I was failing the entire time.

THE PLANT HEALTH PYRAMID, NUTRIENTS REQUIRED, AND MITIGATION OF PESTS

It is reported that most of our ornamental and agricultural crops currently photosynthesize at rate of 10-20% of their full capacity. So just by optimizing the nutrient levels required by the plant - we can significantly increase photosynthesis. The major nutrients required for photosynthesis are magnesium, iron, manganese, nitrogen and phosphorous. Once adequate photosynthesis is taking place, the plant develops greater resistance to soil-borne fungal pathogens such as Verticillium, Fusarium, Rhizoctonia, Pythium, Phytophthora, and others. In most fungal pathogen–plant systems, a high level of sugars in plant tissues enhances plant resistance. https://www.ncbi.nlm.nih.gov/pmc/articles/PMC4219568/

Sugars are also involved in metabolic and signalling pathways of plants. Sugar signals can contribute to immune responses against pathogens – leading to pathogen-associated triggered immunity in plants https://academic.oup.com/jxb/article/63/11/3989/604616

Level two of the plant health pyramid (Fig. 4) entails increasing protein synthesis, utilizing nitrogen compounds and converting amino acids into protein. The major nutrients required for protein synthesis include magnesium, sulfur, molybdenum, and boron. Plants become resistant to insects with simple digestive systems, especially larval and sucking insects such as tomato horn worms, cabbage loopers, corn borers, aphids, leafhoppers, whiteflies, and thrips.

Levels three and four of the plant health pyramid entail active immunity and are dependent on high functional levels of soil biology. Level three is increased lipid synthesis. This provides resistance to airborne fungal and bacterial pathogens such as downy mildew, fire blight, rust, bacterial speck, late blight, etc.

Level four entails increasing plant secondary metabolites. This provides resistance to beetles, nematodes, and viruses. Obviously, this means that the plant health needs to be clicking on all cylinders to achieve resistance to these pests. For those of you, who like us, suffer from red-headed flea beetles, I sympathize! I can say that supplementing molybdenum has helped us somewhat on reducing the population of the red-headed flea beetle. Evidently our plants did not have adequate amounts of molybdenum to synthesize the required amount of nitrate reductase enzyme.

So how do you get all these required minerals into the plant to satisfy levels 1 through 4? All you can do is set the stage with base saturation of the soil/ media, introduce microbes into the system and supplement with foliar feeding.

FOLIAR FERTILIZATION

How many of you have tried foliar feeding, never saw any results, and quickly gave up on it? Through most of my career, I had the same problem until I learned that there are good and bad times to foliar feed. Plants will respond 8-20 times more efficiently to foliar sprays compared to soil applications.

In layman's terms the *point of deliquescence* is simply the point at which a foliar spray (ionic salts) will dry (into crystals) on the leaf surface. It is important to keep the foliar spray in solution on the leaf for as long as possible to aid absorption. This is why evening sprays are ideal as they take advantage of higher humidity at night. Also, urea, magnesium chloride, calcium nitrate and potassium nitrate can be added to the spray to help keep the sprays in solution longer.

FERTILIZER, FUNCTION AND ENHANCED PEST RESISTANCE

Proper use of fertilizers can promote plant growth and enhance pest resistance. Fertilizers include calcium chloride, potassium nitrate, calcium nitrate - but avoid potassium chloride (muriate of potash) if possible, which can be toxic to microbes. The primary nutrients that promote reproductive growth are ammonium, phosphorous and manganese.

Boron increases calcium absorption from the soil and mobility within the plant. Calcium is the bus and boron is the bus driver. Ideal soil ratio is 1:400 Bo: Ca. All crops can benefit from a light supplemental foliar spray of boron prior to flowering. Boron increases resistance to Fusarium, Verticillium, Rhizoctonia, and some viruses.

Foliar applications of the micronutrients molybdenum and cobalt have enhanced our plants. If time-sensitive crops such as poinsettias, mums, etc. are finishing ahead of schedule - a cobalt spray will slow down their growth.

Copper is essential for fruit tree production, provides bark stretchiness and resistance to a host of bacterial and fungal diseases. Cracks in the bark of your trees or rabbit tracks on your crape myrtle leaves are indicative of copper deficiency.

Iron is a major component of multiple enzymes, protein synthesis, chlorophyll and the oxygen carrying system within the plant. It also helps mitigate some of the bacterial diseases.

Manganese also plays a major role with enzyme systems within the plant. Chlorophyll production, carbohydrate and nitrogen metabolism all depend on manganese. Manganese is best applied as a foliar application. Combine manganese with phosphoric acid to aid absorption. Manganese helps mitigate a wide range of fungal and bacterial pathogens.

Molybdenum is required for the formation of nitrate reductase enzyme which converts nitrates into amino acids. Molybdenum helps mitigate a wide range of pathogens from viruses, bacteria, fungi, and nematodes. Blueberries (*Vaccinium*), do not have the ability to make nitrate reductase enzyme, hence molybdenum is less important to them - and nitrate nitrogen can be toxic to them because of this.

Phosphate is the major catalyst in all living systems and important in photosynthesis. The more efficiently plants take up water and nutrients, the higher the photosynthetic levels and carbohydrate/ sugar production - the higher the Brix level. Phosphate is key to obtaining a high brix crop. If there is a phosphate deficiency there is typically a deficiency in all other nutrients. Phosphorous mitigates bacterial, fungal and nematode pests.

Sulfur is useful in reducing cations that are in excess. It bonds with them and makes them water soluble. Sulfur also lowers the soil pH. Magnesium sulfate, calcium sulfate, and potassium sulfate are all very water soluble. Sulfur-induced resistance is a very real phenomenon. Sulfur helps mitigate everything from fungi to spider mites.

TOOLS REQUIRED TO OBTAIN THE NECESSARY INFORMATION NEEDED FOR ACCURATE NUTRITION DECISIONS

All nutrition management decisions are based on irrigation water, soil (media), and leaf tissue analyses. For instance, a soil analysis can help determine how much calcium is needed to reach the proper base saturation levels.

In leaf tissue analysis, younger and older leaves on the plant are separated. Leaf samples are taken from two different locations of a plant, based on maturity, and bagged separately. If it is a mobile element, such as magnesium, and the younger leaves have more magnesium than the older leaves - then the plant probably does not have enough total magnesium because its robbing from the older leaves to transport it to the younger. If both the younger and the lower leaves are on the low end of the spectrum then you need a foliar spray for a faster response. I would spray a tank mix of potassium nitrate, potassium silicate, sea salt, solubor and sulfur.

Remember that we are the original Green Industry…it's time we started acting like it!

Weed Control in Propagation: Hand Weeding is NOT the Only Option

Anthony Witcher and Isha Poudel

Tennessee State University, Otis L. Floyd Nursery Research Center, McMinnville, TN 37110

awitcher@tnstate.edu

Keywords: Mulch, pre-emergent herbicide, seedlings, stem cuttings

Summary

Weed control in nursery crop propagation is difficult due to the limited methods that are safe and effective. Hand weeding is labor intensive and time consuming and the availability of agriculture labor has become limited in recent years. Adoption of sanitation practices helps minimize weed infestations, but utilization of pre-emergent herbicides and mulches may be a viable weed control method in propagation. Although certain pre-emergent herbicides may cause injury to seedlings and rooting cuttings, there are non-root-inhibiting herbicides that may be safe for use in propagation. Three studies were conducted evaluating pre-emergent herbicides and mulches in seedling and stem-cutting propagation in small diameter containers. For seedling propagation, we found that isoxaben (Gallery) was safe when applied to small seedlings of several tree species after transplant and several pre-emergent herbicides were safe when applied prior to germination of oak seeds. For stem cutting propagation, oxadiazon+oxyfluorfen (Regal O-O) provided broad spectrum weed and was safe when applied 2 weeks after sticking cuttings of several crop species. Pine pellet mulch provided excellent weed control at 0.5-inch depth with no impact to cutting root development.

INTRODUCTION

Plant propagation is a key component of the nursery industry, with many nurseries specializing in propagating crops for sale to other nurseries for growing on to a finished size. In the United States, propagative material is produced on 24,192 acres with annual sales of $753 million. The majority of nursery crops are propagated by seed or vegetative cuttings. Although many nursery crop species can be propagated in field beds, container-grown seedlings and rooted cuttings provide advantages such as an extended transplant season and increased transplant success (Fare, 2013). Weeds are a major issue in container-grown propagative material but limited weed control methods are available for use during propagation.

Weeds can become established and produce seed within a few weeks, quickly infesting liners prior to finishing. Although weeds can be removed prior to shipping, seeds have already infested the containers and will be a problem in crop production. Manual weed removal (hand weeding) is the most common method of weed control in propagation, but it is time consuming, costly, and requires a labor pool that has become diminished in recent years. As a result, growers must prioritize tasks which can result in less frequent hand weeding. Development of improved weed control methods is needed to reduce labor and cost inputs during propagation and improve crop quality.

SANITATION

Weeds can infest propagation from a number of sources including container substrate, containers, floors within the propagation area, surrounding areas, stock plants, and workers. Container substrates, especially pine bark in bulk piles, should be stored indoors or in a protected area to prevent weed seed infestation. Propagation containers that are re-used should be thoroughly cleaned with high pressure water sprays to remove weed seeds, especially seeds with a sticky outer coating such as bittercress and woodsorrel (Neal, 2016). Surrounding areas, nearby container production blocks, floors of the propagation space, and stock plants used for cuttings should be maintained weed free to prevent infestation of the crops in propagation.

Post-emergent herbicides can be used to control actively growing weeds, but care must be taken to avoid contact with foliage of desirable crops. Several post-emergent herbicides are labeled for use inside structures (such as greenhouses) and can be used to control weeds during propagation, these products include diquat (Reward), glufosinate (Finale), glyphosate (Round-Up), and pelargonic acid (Scythe). Pre-emergent herbicides can be used to prevent weed seed establishment in container-grown crops in production, on gravel production pads, and non-crop areas such as gravel drives and walkways. Pre-emergent herbicides such as flumioxazin (Sureguard) and indaziflam (Marengo) can also be used on greenhouse floors, but these products should be applied prior to moving in flats/containers.

HERBICIDE USE IN PROPAGATION

Currently, there are no pre-emergent herbicides labeled for use in propagation and many products restrict use in small diameter containers (less than 4 inches) and on non-rooted cuttings. Additionally, no pre-emergent herbicides are labeled for use in enclosed structures. Seed and cutting propagation involves the initiation, growth, and development of new roots which are sensitive to chemical substances such as herbicides. As a result, pre-emergent herbicides have not been widely used during propagation. Pre-emergent herbicides function by inhibiting germination or root/shoot development and sensitivity can vary by chemical class and plant species. Pre-emergent herbicides in the dinitroaniline family act as root inhibitors and numerous research reports noted reduced rooting percentage and root development when used in cutting propagation. Root inhibiting herbicides such as oryzalin (Surflan), pendimethalin (Pendulum), prodiamine (Barricade), and trifluralin (Treflan) should not be used in propagation.

Nevertheless, other studies have shown that certain non-root-inhibiting pre-emergent herbicides could be safely applied during propagation. In seedling propagation, Willoughby et al. (2003) reported that isoxaben and pendimethalin could be applied prior to seed germination of several woody species. South and Carey (2005) found that oxyfluorfen (Goal) was safe to apply to several large-seeded tree species (*Carya* spp, *Juglans* spp., and *Quercus* spp.) prior to germination. Halcomb and Fare (1997) demonstrated that isoxaben did not damage small field-grown tree seedlings when applied over the top of actively growing plants. In cutting propagation, oxadiazon (Ronstar) has been found safe to apply prior to sticking cuttings of several crop species (Johnson and Meade, 1986; Langmaid, 1987; Thetford et al., 1988; Thetford and Gilliam, 1991). Isoxaben was also safe to apply to *Loropetalum chinense* at various stages of propagation (Cochran et al., 2008).

RESEARCH AT TENNESSEE STATE UNIVERSITY

Most of the previous work evaluating pre-emergent herbicides in propagation was completed over twenty years ago and there are newer products that may be viable for weed control in propagation. Pre-emergent herbicides such as flumioxazin and indaziflam may have potential for use during propagation, while other types of products such as mulches may be viable alternatives for weed control in sensitive crops and inside greenhouses. In recent years, several studies have been completed at the Tennessee State University Otis L. Floyd Nursery Research Center in McMinnville, TN evaluating pre-emergent herbicides and mulches in seedling and cutting propagation.

In the first study, container-grown tree seedlings were treated with various pre-emergent herbicides and mulches (Table 1). Containers (3.5-inch diameter filled with pine bark substrate) planted with seeds of two oak species [sawtooth oak (*Quercus acutissima*) and willow oak (*Q. phellos*)] were treated prior to seed germination, while seedlings (128 cell trays) of four other tree species [kousa dogwood (*Cornus kousa*), sweet gum (*Liquidambar styraciflua*), sweetbay magnolia (*Magnolia virginiana*), and yellow poplar (*Liriodendron tulipifera*)] were transplanted to containers and treated after 3 days. Compared to the non-treated control, reduced root dry weight was only observed for kousa dogwood (dimethenamid-P+pendimethalin, pendimethalin, pine pellets, prodiamine, and trifluralin) and yellow poplar (trifluralin) (Figure 1). Weed control efficacy varied by product and weed species but pine pellets provided excellent of bittercress and large crabgrass. Overall, several pre-emergent herbicides were safe and effective for use in seedling propagation of several tree species.

Table 1. Mulches and pre-emergent herbicides evaluated in a seedling propagation trial.

Product Type	Product	Active ingredient
NA	Control	NA
Mulch (0.3-inch depth)	Perlite	NA
	Pine Pellets	NA
	Cedar Shavings	NA
	Charcoal	NA
Herbicide (High Label Rate)	Treflan F	trifluralin
	Treflan 5G	trifluralin
	Pendulum AC	pendimethalin
	Pendulum 2G	pendimethalin
	Gallery SC	isoxaben
	Snapshot 2.5TG	trifluralin + isoxaben
	Barricade 4FL	prodiamine
	Freehand G	dimethenamid-P + pendimethalin

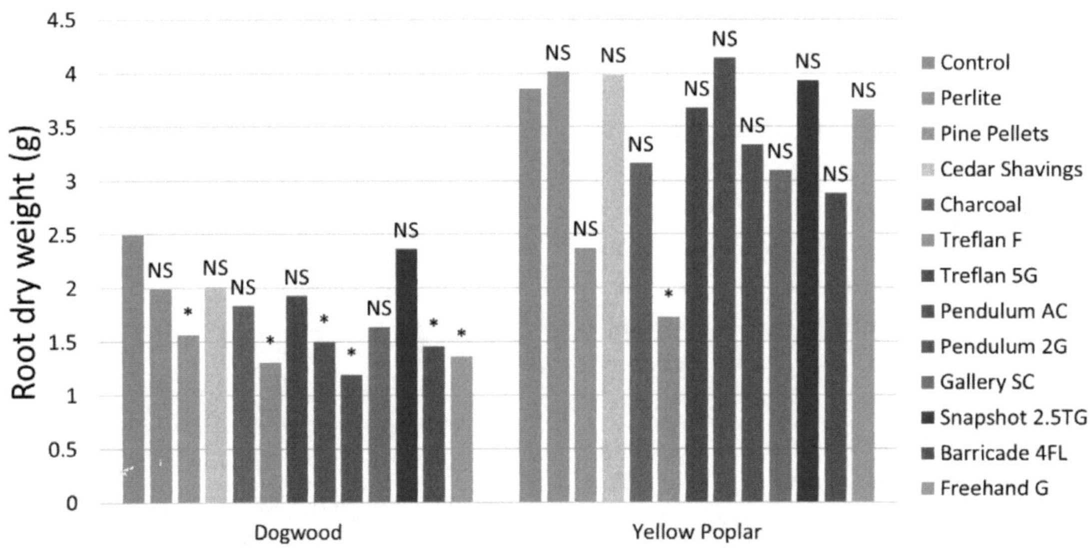

Figure 1. Root dry weight of kousa dogwood and yellow poplar seedlings treated with mulches and pre-emergent herbicides. Compared to the non-treated control, NS = not significant and * = significantly different at $p < 0.05$.

In the second study, stem cuttings (stuck in 2.5-inch diameter containers filled with pine bark substrate) of three crop species [butterfly bush (*Buddleja davidii* 'Nanho Blue', holly (*Ilex cornuta* 'Dwarf Burford'), and viburnum (*Viburnum plicatum f. tomentosum* 'Mariesii')] were treated with various mulches (prior to sticking) and pre-emergent herbicides (2 weeks after sticking) (Table 2). Weed control efficacy was also evaluated for four weed species [bittercress

(*Cardamine hirsuta*), crabgrass (*Digitaria sanguinalis*), creeping woodsorrel (*Oxalis corniculata*), and mulberryweed (*Fatoua villosa*)]. Compared to the non-treated control, rooting percentage was only reduced for butterfly bush with isoxaben while rooting percentage was not affected for holly or viburnum (data not shown). Similarly, butterfly bush root and shoot dry weight was only reduced by isoxaben while no differences were observed for the other crop species (data not shown). Oxadiazon+oxyfluorfen provided excellent control of all tested weed species and has potential for propagation of a number of crops. Mulches did not provide adequate weed control of all weed species, but increased application depth may enhance efficacy.

Table 2. Mulches and pre-emergent herbicides evaluated in a stem cutting propagation trial.

Product Type	Product	Active Ingredient(s)
Non-treated control	NA	NA
Herbicide (High label rate)	Gallery SC	isoxaben
	BroadStar	flumioxazin
	Marengo G	indaziflam
	Regal O-O	oxadiazon+oxyfluorfen
	Ronstar G	oxadiazon
Mulch (0.3 inch depth)	Rice Hulls	NA
	Vermiculite	NA
	Pine Pellets	NA
	Paper Pellets	NA

In the third study, mulch type and depth were evaluated for rooting cuttings of three crop species [butterfly bush (*Buddleja davidii* 'Nanho Blue'), crape myrtle (*Lagerstroemia indica* 'Catawba'), and hydrangea (*Hydrangea paniculata* 'Phantom')]. Mulches included coarse vermiculite, paper pellets, pine pellets, and rice hulls applied at 0.5- or 1-inch depth prior to sticking cuttings (Figure 2). Weed control efficacy was also evaluated for creeping woodsorrel, bittercress, crabgrass, and mulberry weed. No differences in rooting percentage were observed for any treatments. Crape myrtle root dry weight was lower for paper pellets (both depths), but no differences were observed for butterfly bush or hydrangea. Pine pellets and paper pellets (both depths) reduced growth of all four weed species. Pine pellets and paper pellets at 0.5-inch depth can be effective in suppressing the weed population with minimal effects on rooting.

In summary, we demonstrated several pre-emergent herbicides and mulches have potential for use in seedling and cutting propagation. Several pre-emergent herbicides may be applied prior to germination of large-seeded tree species while small seedlings of certain tree species can be safely treated with isoxaben after transplant. For cutting propagation, oxadiazon+oxyfluorfen was safe when applied 2 weeks after sticking cuttings of several crop species and provided broad spectrum weed control under intermittent mist. Paper and pine pellet mulches may be alternatives to pre-emergent herbicides for use on sensitive crop species and in enclosed structures and provided excellent weed control at 0.5-inch depth. Growers should conduct small trials with individual products and crop species prior to large scale adoption.

Figure 2. Paper pellet, pine pellet, and rice hull mulches (left to right) when applied dry (top) then saturated with irrigation (bottom).

LITERATURE CITED

Cochran, D.R., Gilliam, C.H., Eakes, D.J., Wehtje, G.R., and Knight, P.R. (2008). Herbicide use in propagation of *Loropetalum chinense* 'Ruby'. J. Environ. Hortic. *26*:139-143.

Fare, D.C. (2013). Propagation container and timing of propagation affects growth of oak seedlings. J. Environ. Hort. *31*:43-48.

Halcomb, M.A., and Fare, D.C. (1997). Tolerance of seedlings to preemergence herbicides. Proc. South. Nurs. Res. Conf. *42*:315-318.

Johnson, J. R., and Meade, J.A. (1986). Preemergent herbicide effect on the rooting of cuttings. Proc. Intl. Plant Prop. Soc. *36*:567-570.

Langmaid, M. (1987). Herbicides used in propagation at Wight Nurseries. Proc. Intl. Plant Prop. Soc. *37*:539-540.

Neal, J. (2016). Weed control in woody plant propagation and containerized liner production. N. Carolina St. Univ. Coop. Ext. Bul. AG-287.

South, D.B., and Carey, W.A. (2005). Weed control in bareroot hardwood nurseries. USDA Forest Serv. Proc. *35*:34-38.

Thetford, M., and Gilliam, C.H. (1991). Herbicide use in propagation: Effects on rooting and root growth of stem cuttings. J. Environ. Hort. *9*:21-23.

Thetford, M., Gilliam, C.H., and Foster, W.J. (1988). Herbicide use in propagation. Proc. Intl. Plant Prop. Soc. *38*:479-481.

Willoughby, I., Clay, D., and Dixon, F. (2003). The effect of pre-emergent herbicides on germination and early growth of broadleaved species used for direct seeding. Forestry *76*:83-94.

International Plant Propagators' Society (IPPS)
174 Crestview Drive
Bellefonte, PA 16823-8516
USA

ISBN 978-1-7138-4505-8
9 781713 845058

CURRAN ASSOCIATES INC.
proceedings.com